ILLUSTRATED DICTIONARY FOR ELECTRICAL WORKERS

Second Edition

Patricia A. Titus

James E. Titus

John Traister

Australia   Canada   Mexico   Singapore   Spain   United Kingdom   United States

Illustrated Dictionary for Electrical Workers, 2nd Edition
by
Patricia A. Titus, James E. Titus, and John Traister

Business Unit Director:
Alar Elken

Executive Editor:
Sandy Clark

Acquisitions Editor:
Mark Huth

Development:
Dawn Daugherty

Executive Production Manager:
Mary Ellen Black

Production Editor:
Ruth Fisher

Executive Marketing Manager:
Maura Theriault

Marketing Coordinator:
Brian McGrath

COPYRIGHT © 2002 by Delmar, a division of Thomson Learning, Inc. Thomson Learning™ is a trademark used herein under license

Printed in Canada

1 2 3 4 5 XXX 05 04 03 02 01

For more information contact Delmar,
3 Columbia Circle, PO Box 15015,
Albany, NY 12212-5015.
Or find us on the World Wide Web at http://www.delmar.com

ALL RIGHTS RESERVED. No part of this work covered by the copyright hereon may be reproduced or used in any form or by any means—graphic, electronic, or mechanical, including photocopying, recording, taping, Web distribution or information storage and retrieval systems—without written permission of the publisher.

For permission to use material from this text or product, contact us by
Tel (800) 730-2214
Fax (800) 730-2215
www.thomsonrights.com

Library of Congress Cataloging-in-Publication Data

Titus, Patricia A.
    Illustrated dictionary for electrical workers / Patricia A. Titus, James E. Titus, John Traister.— 2nd ed
      p. cm.
    First ed. by John Traister.
    ISBN 0-7668-2853-0 (alk. paper)
    1. Electrical engineering—Dictionaries.
I. Titus, Patricia A. II. Titus, James E.
III. Traister, John. Illustrated dictionary for electrical workers. IV. Title.
TK9.T58   2002
621.3'03—dc21   2001028183

NOTICE TO THE READER

Publisher does not warrant or guarantee any of the products described herein or perform any independent analysis in connection with any of the product information contained herein. Publisher does not assume, and expressly disclaims, any obligation to obtain and include information other than that provided to it by the manufacturer.

The reader is expressly warned to consider and adopt all safety precautions that might be indicated by the activities herein and to avoid all potential hazards. By following the instructions contained herein, the reader willingly assumes all risks in connection with such instructions.

The Publisher makes no representation or warranties of any kind, including but not limited to, the warranties of fitness for particular purpose or merchantability, nor are any such representations implied with respect to the material set forth herein, and the publisher takes no responsibility with respect to such material. The publisher shall not be liable for any special, consequential, or exemplary damages resulting, in whole or part, from the readers' use of, or reliance upon, this material.

Preface

Because of the rapidly changing world of electricity and electronics, the *Illustrated Dictionary for Electrical Workers*, Second Edition, has been enlarged and updated to assist you as a student in college, technical school, or electrical apprenticeship training program. Professionals, such as electrical designers, fiber optics technicians, journeyman electricians, maintenance electricians, and other electrical workers, should also find this dictionary invaluable.

The *Illustrated Dictionary for Electrical Workers*, Second Edition, is especially important because many textbooks have a glossary, but very few, if any, cover the wide range of terminology necessary to have a well-rounded understanding of the many aspects of the electrical and electronics fields. And, as you have probably already found out, definitions for many electrical terms are not found in the average dictionary or in any technical text, yet these terms are used extensively in the electrical field throughout the United States and Canada. With almost 2000 terms and about 200 illustrations, the *Illustrated Dictionary for Electrical Workers*, Second Edition, becomes an invaluable tool for meeting these needs.

We have written this book in the hope that it will assist you in the pursuit of excellence in your field.

Pat and Jim Titus
Alamo, Texas

Acknowledgments

The authors and Delmar wish to thank the members of the review panel for their many suggestions and comments. A hearty thanks to:

Gary Reiman
Dunwoody Technical Institute
Minneapolis, MN

Madeline Borthick
Independent Electrical Contractors
Houston, TX

Cliff Rediger
Independent Electrical Contractors
Denver, CO

AA (Aluminum Association): A manufacturers' association that promotes the use of aluminum.

AAC: All Aluminum Conductor.

AASC: Aluminum Alloy Stranded Conductors.

abrasion: The process of rubbing, grinding, or wearing away by friction.

abrasion resistance: Ability to resist surface wear.

abrasive paper: Paper or cloth on which flint, garnet, emery, aluminum oxide, or corundum has been fastened with glue or some other adhesive. One use in the electrical field is to clean conductors, contacts, or terminals.

abscissa: In an ordered pair the first number, or x-coordinate, is the abscissa.

ac (alternating current): 1) A periodic current, the average of which is zero over a period; normally the current reverses after given time intervals and has alternately positive and negative values. 2) The type of electrical current actually produced in a rotating generator (alternator).

accelerated life tests: Subjecting a product to operating conditions more severe than normal to expedite deterioration, affording some measure of probable life at normal conditions.

accelerating relay: A type of relay that assists in starting or accelerating a motor to different speeds.

accelerator: 1) A substance that increases the speed of a chemical reaction. 2) Something to increase velocity.

accent lighting: Fixtures that provide a lighted focal point on an area or object.

acceptable (nuclear power): Demonstrated to be adequate by the safety analysis of the station.

acceptance test: Made to demonstrate the degree of compliance with specified requirements.

accepted: Approval for a specific installation or arrangement of equipment or materials.

accessible: Capable of being removed or exposed without damaging the building structure or finish, or not permanently closed in by the structure or finish of the building.

across-the-line: A means of starting a motor in which the motor is connected directly to the line; this method is used for either starting or running.

ACSR (Aluminum, Conductor, Steel Reinforced): A bare composite of aluminum and steel wires, usually aluminum around steel.

actuated equipment (nuclear power): Component(s) that perform a protective function.

actuator: The mechanical part of a limit switch that transfers the mechanical force to the electrical contacts in order to stop or reverse a motor when certain limits have been reached.

acute angle: Any angle that is less than a right angle.

addendum: A modification or addition to the bid documents, plans, and/or specifications.

additive polarity testing: A test in which one lead of the primary transformer is connected to the secondary lead and the voltage is measured across both windings. If the voltage across the windings equals the sum of the two voltages then what is known as "additive polarity" (boost) exists.

adjacent side: The side that is next to the reference angle that helps form the angle. The adjacent side and the hypotenuse are the two sides that always make up the reference angle. See Fig. A-1.

adjustable bar hanger: A hanger that is designed to connect to studs and/or joists so that electrical boxes can be attached to them. Bar length is adjustable from 10½ to 26½ inches. See Fig. A-2.

adjustable wrench: A hand tool with one adjustable jaw and one fixed jaw designed to turn bolts and nuts (crescent wrench).

adjustment factor: A multiplier found in Table 310-15(b) (2) of the NEC® that is used when more than three current-carrying conductors are in a raceway or cable. (Also called a derating factor.)

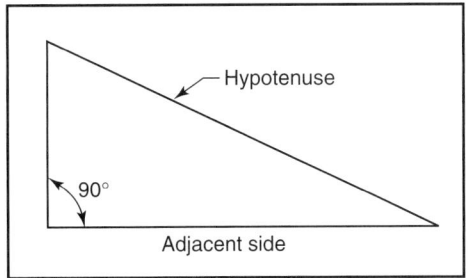

Fig. A-1: Adjacent side of a right triangle and the hypotenuse side.

Fig. A-2: Adjustable bar hanger used between joists to hold electrical boxes.

admedium base: Screw-in type base for mercury reflector lamps.

administrative authority: An organization exercising jurisdiction over the National Electrical Safety Code.

AEC (Atomic Energy Commission): Now defunct; see *ERDA* and *NRC*.

AEIC: Association of Edison Illuminating Companies.

aggregate: Material mixed with cement and water to produce concrete.

aging: The irreversible change of material properties after exposure to an environment for an interval of time.

AIA: 1) American Institute of Architects. 2) Aircraft Industries Association.

AIEE: American Institute of Electrical Engineers.

air cleaner: Device used for removal of airborne impurities.

air diffuser: Air distribution outlet designed to direct airflow into desired patterns.

air entrained concrete: Concrete in which a small amount of air is trapped by addition of a special material to produce greater durability.

air oven: A lab oven used to heat by convection of hot air.

Al: Aluminum.

alarm notification appliance: The sound and/or light signaling part of a smoke detector that will alert occupants to the presence of smoke or fire.

ALCI: Appliance Leakage Circuit Interrupter. Trips at six milliamps and is very useful in older homes that do not have ground-fault circuit-interruptor (GFCI) protection.

Al-Cu: An abbreviation for aluminum and copper, commonly marked on terminals, lugs, and other electrical connectors to indicate that the device is suitable for use with either aluminum conductors or copper conductors.

alive: Energized; having voltage applied.

Allen wrench: A hex-shaped hardened steel rod with a 90° angle that forms a handle. See Fig. A-3.

alligator wrench: A wrench with toothed V-shaped jaws fixed in position.

alloy: A substance having metallic properties and being composed of elemental metal and one or more chemical elements.

Fig. A-3: Allen wrench. Allen wrenches are manufactured in many different sizes.

alternation: A half cycle; the part of the cycle above or below the horizontal line.

alternator: A device to produce alternating current. Alternators range in size from small automotive types (Fig. A-4) to huge types used in power plants to produce electricity for cross-country distribution, as shown in Fig. A-5.

Alumoweld®: An aluminum-clad steel wire by Copperweld Steel Corp.

ambient temperature: Temperature of fluid (usually air) that surrounds an object on all sides.

American bond: Brickwork pattern consisting of five courses of stretchers followed by one bonding course of headers.

ammeter: An electric meter used to measure current, calibrated in amperes. See Fig. A-6.

ampacity: The current-carrying capacity of conductors or equipment, expressed in amperes.

ampere (A): The basic SI unit measuring the quantity of electricity.

ampere-turn: The product of amperes times the number of turns in a coil.

amplification: Procedure of expanding the strength of a signal.

amplifier: 1) A device that enables an input signal to directly control a larger energy flow. 2) The process of increasing the strength of an input.

Fig. A-4: Automotive alternators are used to keep the car's storage battery charged.

amplitude: The maximum value of a wave.

analog: Pertaining to data from continuously varying physical quantities.

AND gate: A device that must have all of its inputs high to produce a high output. See Fig. A-7.

angle: The figure formed by two lines merging at a common point.

angle bracket: A form of support having two faces generally at right angles to each other. A web is often added to increase strength.

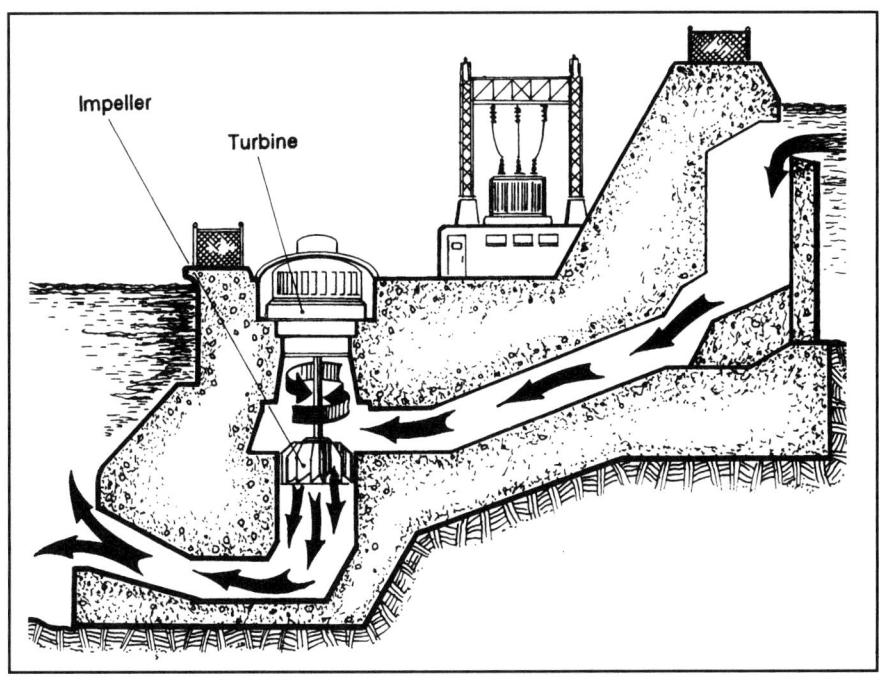

Fig. A-5: Large hydroelectric power plants supply huge amounts of our electrical energy.

Fig. A-6: One type of ammeter (amp clamp) used to read amperage in a conductor.

angle pull: In rigging, the load-carrying capacity of a rope is estimated by using the K-factor as the angle of pull increases. See Fig. A-8.

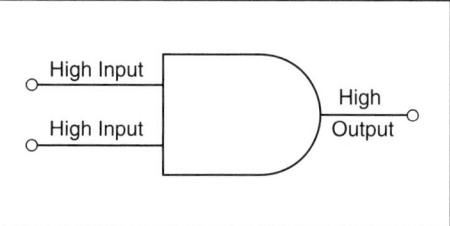

Fig. A-7: AND gate symbol. (A logic symbol used in electronics.)

angle, roll over (overhead): The sum of the vertical angles between the conductor and the horizontal on both sides of the traveler; excessive roll over angles can cause premature splice failures.

angular motion: Movement in a circular motion. Angular motion is usually measured along the circumference of a circle.

angular velocity: The average time rate of change of angular position; in electrical circuits = 2f, and f equals frequency.

ANI (American Nuclear Insurers): A voluntary unincorporated association of companies providing property and liability insurance for U.S. nuclear power plants; formerly *NELPIA*.

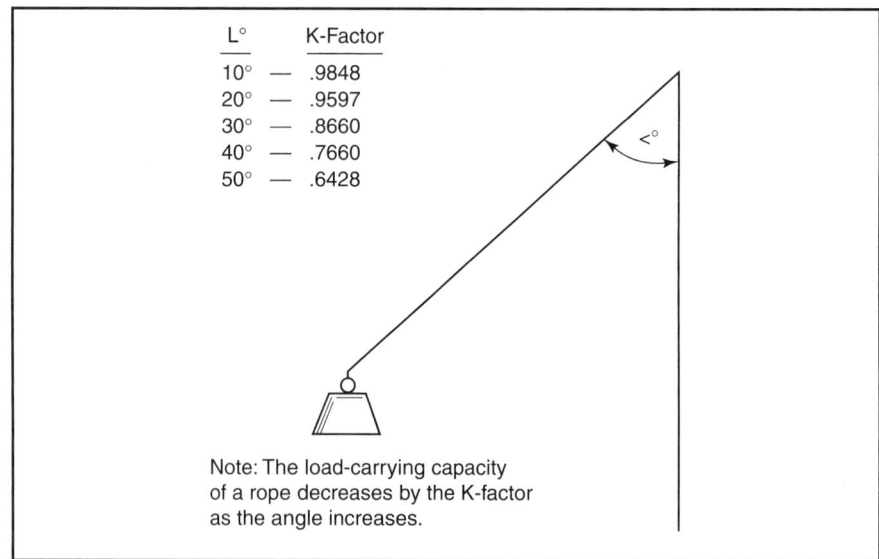

Fig. A-8: Angle pull. A method of estimating the load-carrying capacity of a rope.

annealing: The process of preventing or removing objectional stresses in materials by controlled cooling from a heated state; measured by tensile strength.

annealing, bright: Annealing in a protective environment to prevent discoloration of the surface.

annunciator: A signaling device that is operated electromagnetically.

annunciator wire: Wire used for low-voltage bell, thermostat, and chime circuits.

anode: 1) Positive electrode through which current enters a nonmetallic conductor such as an electrolytic cell. 2) The negative pole of a storage battery.

ANSI (American National Standards Institute): An organization that publishes nationally recognized standards.

antenna: A device for transmission or reception of electromagnetic waves.

antioxidant: Retards or prevents degradation of materials exposed to oxygen (air) or peroxides.

antisiphon trap: Trap in a drainage system designed to preserve a water seal by defeating siphonage.

aperture seal (nuclear): A seal between containment aperture and the electrical penetration assembly.

apparent power: Found by multiplying current times voltage in a circuit containing reactance.

appliance: Equipment designed for a particular purpose, using electricity to produce heat, light, and mechanical motion, etc.; usually complete in itself, generally other than industrial use, normally in standard sizes or types.

appliance information plate: A plate connected to an appliance indicating at least the model number, serial number, watts, and amps. See Fig. A-9.

applied voltage: Total voltage connected to a circuit or device.

approved: 1) Acceptable to the authority having legal enforcement. 2) Per *OSHA*: A product that has been tested to standards and

Fig. A-9: Appliance information plate with important information.

found suitable for general application, subject to limitations outlined in the nationally recognized testing laboratory's listing.

apron: Piece of horizontal wood trim under the sill of the interior casing of a window.

AR: Ampere Rating.

arc: A flow of current across an insulating medium.

arc chute: A device that employs the deionization principle to confine and extinguish the arc for each set of contacts.

Arc-fault circuit interrupter: A device intended to provide protection from the effects of rapid current fluctuations due to arcing conditions.

arc furnace: Heats by heavy current flow through the material to be heated.

arcing damage: Damage caused to electrical equipment because of an electrical arc. Arcing damage is normally thought of as being caused by short circuits or ground faults, but damage is also sometimes caused when contacts are opened and closed.

arcing time: The time elapsing from the severance of the circuit to the final interruption of current flow.

arc resistance: The time required for an arc to establish a conductive path in or across a material.

Fig. A-10: The armature is a moving part in all electric motors and generators.

arc suppressor: A device that dissipates the energy present across contacts when they are opened.

areaway: Open space below the ground level immediately outside a building. It is enclosed by substantial walls.

armature: 1) Rotating machine: the member in which alternating voltage is generated. 2) Electromagnet: the member that is moved by magnetic force. See Fig. A-10.

armor: Mechanical protector for cables; usually a helical winding of metal tape formed so that each convolution locks mechanically upon the previous one (interlocked armor); may be a formed metal tube or a helical wrap of wires. See Fig. A-11.

armor clamp: A connector designed to secure armored cable to an electrical box or other device. See Fig. B-16.

armored cable: (Type AC) Cable with a flexible metal sheathing containing at least two conductors. It is also referred to as BX cable. See Fig. A-11 and B-16.

arrester: Wire screen secured to the top of an incinerator to confine sparks and other products of burning.

ASA: American Standards Association.

Fig. A-11: Type AC (armor-clad) cable. The interlocking metal covering is designed to protect the cable conductors.

as-built plans: A set of plans showing all the changes and corrections that were made during the construction process so that when the project is completed the owner will have a full set of plans that are correct.

ashlar: Squared and dressed stones used for facing a masonry wall; short upright wood pieces extending from the attic floor to the rafters forming a dwarf wall.

askarel: A synthetic insulating oil that is nonflammable but very toxic—being replaced by silicone oils.

ASME: American Society of Mechanical Engineers.

associated circuits (nuclear power): Nonclass 1E circuits that share power supplies or are not physically separated from Class 1E circuits.

astable mode: The mode in which an oscillator is able to continually turn on and off or change from positive to negative output.

ASTM (American Society for Testing and Materials): A group writing standards for testing materials and specifications for materials.

asymmetrical: Not identical on both sides of a central line; unsymmetrical.

atom: According to our present understanding, the atom is believed to consist of a central nucleus composed of protons and neutrons, surrounded by orbiting electrons, as shown in Fig. A-12. The nucleus is relatively large when compared with the orbiting electrons, the same as our sun is large when compared to its orbiting planets.

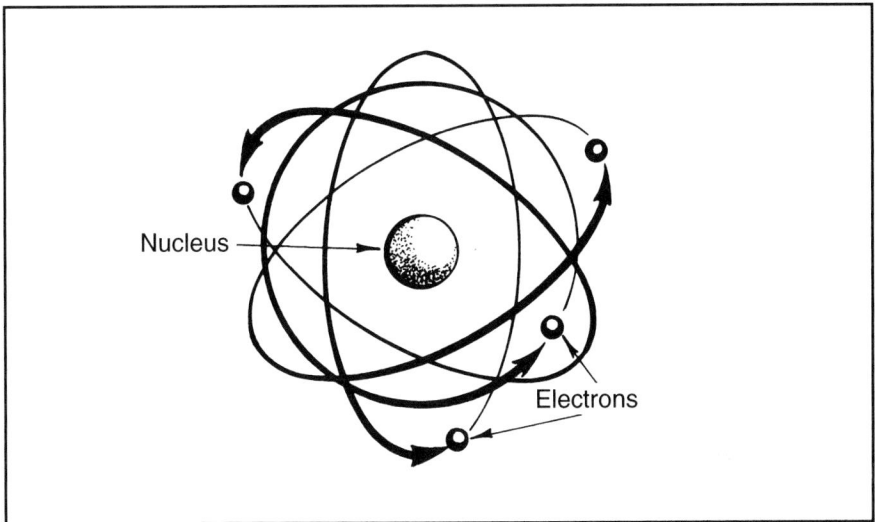

Fig. A-12: An atom is believed to consist of a central nucleus composed of protons and neutrons, surrounded by orbiting electrons.

The orbiting electrons are held in place by the attractive electric force between the electron and the nucleus—similar to how the earth's gravity keeps its satellite (the moon) from drifting off into space. The law of charges states that opposite charges attract while like charges repel. The positive charge of the protons in the nucleus, therefore, attracts the negatively-charges electrons.

If this force of attraction were the only one in force, the electrons would be pulled closer and closer to the nucleus and eventually be absorbed into the nucleus. However, this force of attraction is balanced by the centrifugal force that results from the motion of the electrons around the nucleus.

attachment plug or cap: The male connector for electrical cords. Fig A-13 gives the most common types of receptacles for use with various types of plugs.

attenuation: A decrease in energy magnitude during transmission.

attenuator: A device that reduces the amount of signal voltage or current. Also used in fiber optics, inducing loss by reducing signal power.

Fig. A-13: Receptacle configurations.

attraction: (electrostatic) A force that draws electrically charged objects together.

audible: Capable of being heard by humans.

auditable data: Technical information that is documented and organized to be readily understandable and traceable to independently verify inferences or conclusions based on these records.

auger: A wood-boring tool of large size with handle attached at right angles to the tool line. Several types are made for different purposes.

authority having jurisdiction (AHJ): A government body that has legal jurisdiction over electrical installations.

autoclave: A heated pressure vessel used to bond, cure, seal, or used for environmental testing.

automatic: Operating by own mechanism when actuated by some impersonal influence; nonmanual; self-acting.

automatic starter: A starter designed to control the acceleration of a motor automatically during the acceleration process.

automatic transfer equipment: A device to transfer a load from one power source to another, usually from normal to emergency source and back.

autotransformer: Any transformer where primary and secondary connections are made to a single cell. The application of an autotransformer is a good choice for some users where a 480Y/277- or 208Y/120-volt, three-phase, four-wire distribution system is utilized.

auxiliary: A device or equipment that aids the main device or equipment.

auxiliary contact: Subordinate contacts that operate with the movement of the main contacts.

auxiliary fire alarm system: A warning system connected to a municipal fire alarm system that transmits an alarm to a public fire service communication center.

auxiliary trip relay: An output device attached to an initiating device. It is only activated when its associated initiating device is in the alarm state.

average rated life: The median time it takes for a lamp to burn out. Testing is generally done in the manufacturer's laboratories.

AWG (American Wire Gauge): The standard for measuring wires in America.

awl: A small pointed tool for making holes for nails or screws. When used to mark metal objects, it is sometimes called "scratch awl."

B

backfill: Loose earth placed outside foundation walls for filling and grading; fill of broken stone or other coarse material outside a foundation or basement wall to provide drainage.

back pressure: Pressure in the low side of a refrigerating system; also called suction pressure or low side pressure.

back-to-back bend: A U- shaped bend shaped by two parallel 90° bends with a straight section of pipe between the bends. See Fig. B-1.

baffle: Generally, an opaque shield that directs light in a certain direction. Coves and valances are popular types of baffles.

balanced circuit: An electrical circuit so adjusted with respect to nearby circuits as to escape the influence of mutual induction; a three-wire circuit having the same load on each side of the neutral wire.

ballast: A device designed to stabilize current flow. See Fig. B-2.

balloon framing: System of small house framing; two by fours extending two stories with inch by quarter ledger strips notched into the studs to support the second-story floor beams.

bandwidth: 1) The range of signal frequencies through which a fiber optic device, network, or link will operate. 2) The range

Fig. B-1: Back-to-back bends. Examples of "U"-shaped and "S"-shaped conduit bends.

Fig. B-2: A ballast stabilizes the electric current to provide proper fluorescent operations.

between the low and high frequencies in an *LC* circuit responds within 70.7% of its maximum response.

bank: An installed grouping of a number of units of the same type of electrical equipment; such as "a bank of transformers" or "a bank of capacitors" or a "meter bank," etc.

bar: A long, solid product having one cross-sectional dimension of 0.375 inch or more.

bare (conductor): Not insulated; not coated.

bargeboard: Ornamented board covering the roof boards and projecting over the slope.

bar magnet: 1) A straight permanent magnet. 2) A permanent magnet made in a bar shape.

barometer: Instrument for measuring atmospheric pressure.

barrier: A partition, such as an insulating board to separate bus bars of opposite voltages.

barrier charge: The potential realized across a junction of a semiconductor.

base: The area between the collector and emitter of a transistor. It controls the flow of current through the emitter-base. See Fig. B-3.

base ambient temperature: The temperature of a cable group when there is no load on any cable of the group or on the duct bank containing the group.

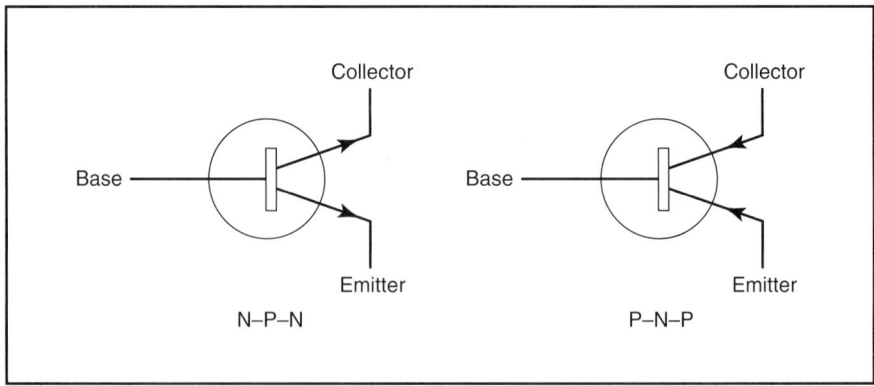

Fig. B-3: Bases for N–P–N and P–N–P transistors.

baseboard: The finishing board that covers the plaster wall where it meets the floor.

baseboard heater: A type of electric heater that is installed at the baseboard level in a room. It may be controlled either by an integral thermostat or one installed elsewhere in the room.

base current: The current that flows through the base-emitter section of a transistor.

base line: A definite line from which measurements are taken in the layout of work.

base load: The minimum load over a period of time.

base speed: The running speed of a DC motor when full-line voltage is applied to its armature and field.

bathroom/heat/vent/light: A ceiling-mounted assembly that is typically used in a bathroom to provide the combined functions of a room heater, a light, and an exhaust fan. See Fig. B-4.

batten: Narrow wood strips used to cover joints.

batter: Slope of the exposed face of a retaining wall.

battery: A device that converts chemical to electrical energy; used to store electricity. See Fig. B-5.

battery acid: The acid used in forming the electrolyte of a storage battery. Sulphuric acid is commonly used.

battery capacity: The number of ampere hours that can be obtained from a storage battery.

battery charger: An electrical device for charging a battery. It usually converts alternating current to direct current for charging purposes.

battery container: The hard rubber or plastic composition casing into which the elements and the electrolyte are placed.

battery resistance: The internal resistance that is present between the plates and the electrolyte in a cell or battery.

bayonet base: Small plug-in base for decorative lamps and exit sign lamps.

bayonet socket: A lamp socket having two lengthwise slots in its sides. These slots take a right-angled turn at the bottom, so that

Fig. B-4: Bathroom light, fan, heater. Fan works with heater or alone.

a lamp with two pins may be connected by pushing it into the socket and giving it a slight turn. See Fig. B-6.

bead: Narrow projecting molding with a rounded surface; or in plastering, a metal strip embedded in plaster at the projecting corner of a wall.

beam: A horizontal member of wood, reinforced concrete, steel, or other material used to span the space between posts, columns, girders, or over an opening in a wall.

bearing plate: Steel plate placed under one end of a beam or truss for load distribution.

bearing wall: Wall supporting a load other than its own weight.

Fig. B-5: Typical storage battery.

bed: Place or material in which stone or brick is laid; horizontal surface of positioned stone; lower surface of brick.

bedding: A layer of material to act as a cushion or interconnection between two elements of a device, such as the jute layer between the sheath and wire armor in a submarine cable; sometimes incorrectly used to refer to extruded insulation shields.

bell transformer: A transformer that changes the line voltage to a useable voltage for door bells and chimes.

bell wire: A type of wire used for low-voltage bell, thermostats, and chime circuits.

belt: The outer protective nonmetallic covering of cable; its jacket.

belted-type cable: A multiple conductor cable having a layer of insulation over the core conductor assembly.

bench mark: Point of reference from which measurements are made as it relates to construction drawings and blueprints.

Fig. B-6: Three bayonet lamp sockets connected in parallel.

Bessel function: A mathematical solution to a differential equation that is used to solve changes in conductor resistance and mutual inductance between conductors with respect to frequency changes due to skin and proximity effects.

bias: The DC voltage that is applied to the base of a transistor to establish its operation point.

bias vacuum tube: Difference of potential between control grid and cathode. Transistor—difference of potential between base and emitter and base and collector. Magnetic amplifier—level of flux density in the magnetic amplifier core under no-signal condition.

bid: A proposal to furnish supplies or equipment; to carry out or to perform certain work for a specified sum.

BIL (Basic Impulse Level): A reference impulse insulation strength.

bimetal strip: Temperature regulating or indicating device that works on the principle that two dissimilar metals with unequal expansion rates, welded together, will bend as temperature changes.

binary coded decimal (BCD): A system that uses four binary digits that represents the numerals 0 through 9. Each of the digits have a weighted value.

binder: Material used to hold assembly together.

bi-pin base: Fluorescent tube plug-in base with two pins. See Fig. B-7.

bipolar: Having two magnetic poles of opposite polarity.

bi-post base: A two-post plug-in base for planar filament lamps generally used for film and TV studio spotlights.

birdcage: The undesired unwinding of a stranded cable.

bit: A boring tool that is used in drill motors, push-pull drills, hand drills, brace and bits, or hand drills. The common purpose is to drill clean holes in wood, metal, concrete, or other materials.

blackbody: A hypothetical body that absorbs, without reflection, all of the electromagnetic radiation incident on its surface.

Fig. B-7: Bi-pin base. A fluorescent tube with two pins.

bleeder resistor: A resistor found in a voltage divider circuit or power supply. It is used to stabilize the output voltage.

blister: A defect in metal, on or near the surface, resulting from the expansion of gas in the subsurface zone. Very small blisters may be called "pinheads" or "pepper blisters."

block (rigging): Consists of wood or metal insides that are fitted with one or more sheaves (pulleys). Size of block depends on size of rope to be used. The block should be about three times the circumference of the rope. See Fig. B-8.

block bridging: Solid wood members nailed between joists to stiffen a floor.

blowout coil: An electromagnetic coil built into a contactor or starter to deflect the arc.

blueprint: See *drawing, electrical.*

Fig. B-8: Block. Wood or metal sheaves (pulleys) used for rigging.

bobbins: Metal spools; small metal reels.

boiler: Closed container in which a liquid may be heated and vaporized. See Fig. B-9.

boiling point: Temperature at which a liquid boils or generates bubbles of vapor when heated.

bond: A mechanical connection between metallic parts of an electrical system, with a bonding locknut or bushing, such as between a neutral wire and a meter enclosure or service conduit to the

Fig. B-9: Electric boiler. Hot water or steam is created by electrical heating coils.

service-equipment enclosure; the junction of welded parts; the adhesive for abrasive grains in grinding wheels.

bonding bushing: A special conduit bushing equipped with a conductor terminal to take a bonding jumper; also has a screw or other sharp device to bite into the enclosure wall to bond the conduit to the enclosure without a jumper when there are no concentric knockouts left in the wall of the enclosure.

bonding grid: Electrically ties all metal parts of a structure, like a pool or metal building installation, together.

bonding jumper: A bare or insulated conductor used to ensure the required electrical conductivity between metal parts required to be electrically connected. Frequently used from a bonding bushing to the service equipment enclosure to provide a path around concentric knockouts in an enclosure wall; also used to bond one raceway to another.

bonding locknut: A threaded locknut for use on the end of a conduit terminal, but a locknut equipped with a screw through its lip. When the locknut is installed, the screw is tightened so its end bites into the wall of the enclosure close to the edge of the knockout. See Fig. B-10.

booster: A generator connected in series with a circuit for the purpose of increasing the voltage of that circuit. Generally used in connection with a system where a storage battery carries part of the load. The booster increases the voltage to a point necessary for charging the battery.

bounceless switch: A circuit designed to expunge contact bounce in mechanical contacts. Generally a digital logic gate is the type of switch constructed.

box (outlet, device, pull, junction, conduit): Used to provide the essential space to

Fig. B-10: Panelboard housing ("can") utilizing bonding locknuts and jumpers.

make electrical connections, accommodate conductors and/or support lighting fixtures.

box fill: The maximum number of conductors and/or devices permitted to be installed in an electrical box or enclosure.

brace: A crank-like handle with an adjustable chuck that holds bits and is used to bore holes in wood.

braid: An interwoven cylindrical covering of fiber or wire.

braking: Any number of ways to stop a motor other than coasting.

branch circuit: That portion of a wiring system extending beyond the final overcurrent device protecting a circuit.

branch circuit center: A central supply where the final overcurrent devices are located and where the branch circuits extend to the outlets.

branch splice: The connection where a wire or conductor taps off from another wire or conductor.

braze: The joining together of two metal pieces, without melting them, using heat and diffusion of a jointing alloy of capillary thickness.

breadboard: Laboratory idiom for an experimental circuit.

breakdown: The abrupt change of resistance from high to low, allowing current flow; an initial rolling or drawing operation.

breakdown torque (BDT): Effect when torquing force is depleted and the motor is brought to a stop or is stalled by the load.

breaker strip: Thin strips of material placed between phase conductors and the grounding conductor in flat parallel portable cables; the breaker strips provide extra mechanical and electrical protection.

breakout: The point at which conductor(s) are taken out of a multiconductor assembly.

bridge: A circuit that measures by balancing four impedances through which the same current flows:
- Wheatstone—resistance
- Kelvin—low resistance
- Schering—capacitance, dissipation factor, dielectric constant
- Wien—capacitance, dissipation factor

bridge circuit: Pressure sensors are designed to produce an output voltage or current that depends on the amount of pressure being applied. The sensor or chip contains four piezoresistors connected to form a bridge.

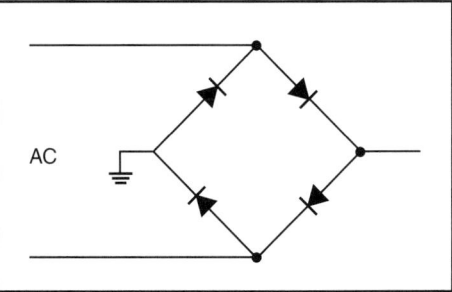

bridge rectifier: A type of full-wave rectifier that changes alternating current into pulsating current. See Fig. B-11.

Fig. B-11: Bridge rectifier. It changes alternating current into pulsating current.

British thermal unit (Btu): Quantity of heat required to raise the temperature of 1 pound of water 1 degree Fahrenheit.

brush: A conductor between the stationary and rotating parts of a machine, usually of carbon. See Fig. B-12.

Fig. B-12: Basic parts of a generator including the carbon brushes.

brush holders: Adjustable arms for holding the commutator brushes of a generator against the commutator, feeding them forward to maintain proper contact as they wear, and permitting them to be lifted from contact when necessary.

brush loss: The loss in watts due to the resistance of the brush contact against the surface of the commutator.

buck: Rough wood door frame placed on a wall partition to which the door moldings are attached; completely fabricated steel door frame set in a wall or partition to receive the door.

buff: To lightly abrade.

bug: A crimped or bolted type of electrical connector for splicing wires or cables together. Also used as a verb: "bugged." Example: The wires were bugged together.

builder's level: An adjustable level (transit) mounted on a tripod and often used in the electrical industry to determine grade levels of finished floors, etc. See Fig. B-13.

Fig. B-13: Builder's level (also called a transit) is an adjustable level mounted on a tripod.

Fig. B-14: Bull-cutters are used for cutting the larger sizes of wire and cable.

building: A separate and distinct structure.

bull-cutters: A large, long-handled tool for cutting the larger sizes of wire and cable, up to kcmil sizes. See Fig. B-14.

buried cable: An electrical or fiber optic cable that is designed for and installed below grade.

bus: The conductor(s) serving as a common connection for two or more circuits. The neutral bus in a circuit breaker panel is shown in Fig. B-15. All branch-circuit neutral conductors terminate here. All ungrounded conductors terminate at the "hot" bus bars.

bus bars: The conductive bars used as the main current supplying elements of panelboards or switchboards; also the conductive bars duct: an assembly of bus bars within an enclosure that is designed for ease of installation, has no fixed electrical characteristics, and allows power to

Fig. B-15: Circuit breaker panelboard showing bus bars, protected by a main circuit breaker, for ungrounded conductors and a neutral bus for grounded and bonding conductors.

be taken off conveniently, usually without circuit interruption. See Fig. B-15.

bushing: A metal, fiber, or plastic fitting designed to screw onto the ends of conduit and cable connectors to provide protection for conductors. Some metal bushings have provisions for bonding jumpers as shown in Fig. B-10.

busway: A metal-enclosed distribution system of bus bars. It is available in prefabricated sections.

butane: Liquid hydrocarbon commonly used as fuel for heating purposes.

buttress: Projecting structure built against a wall to give it greater strength.

butt welding: A weld in which the two pieces to be connected do not overlap but are welded directly at their ends; a common method of welding rods by an electric process.

buzzer: An electric call signal that makes a buzzing noise caused by the rapid vibrations of the armature. It operates on the same principle as the vibrating bell.

BWR (Boiling Water Reactor-nuclear power): A basic nuclear power fission reactor in which steam is used to transfer the energy from the reactor.

BX: A nickname for type AC armored cable (wires with a spiral-wound, flexible steel outer jacketing); although used generically, BX is a registered trade-name of the General Electric Company. See Fig. B-16.

bypass: Passage at one side of or around a regular passage. A bypass switch, for example, would route current around a device or machine.

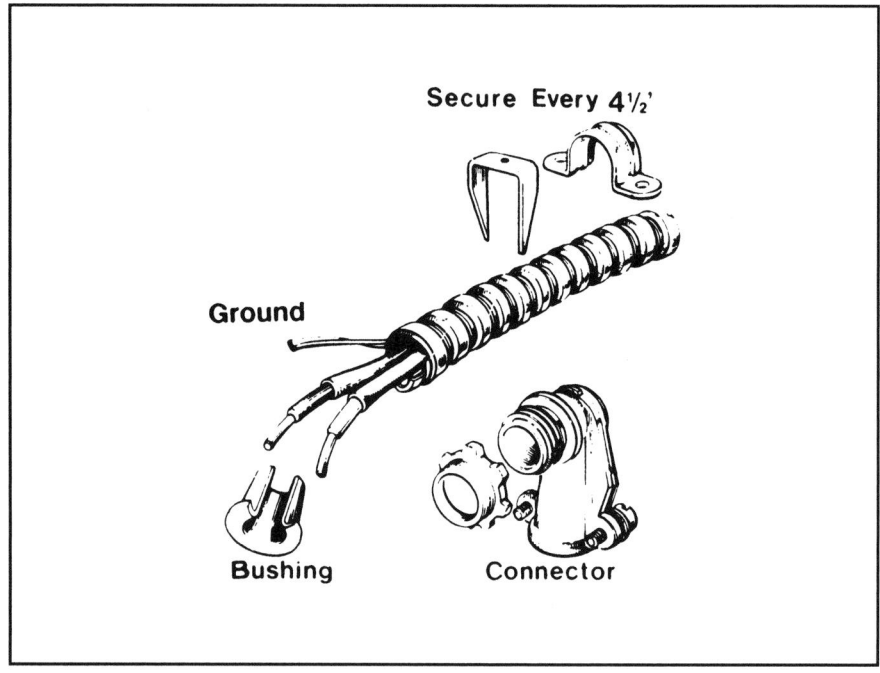

Fig. B-16: Type AC (armored or "BX") cable with insulating bushing and approved fasteners.

C

cabinet (electrical): A housing that encloses branch circuit connections and overcurrent protection devices.

cable: An assembly of two or more wires that may be insulated or bare.

cable, aerial: An assembly of one or more conductors and a supporting messenger. Sometimes referred to as "messenger cable." See Fig. C-1.

cable, armored: A cable having armor (see *armor*).

cable, belted: A multiconductor cable having a layer of insulation over the assembled insulated conductors.

cable, bore-hole: The term given vertical-riser cables in mines.

cable clamp: A device used to clamp around a cable to transmit mechanical strain to all elements of the cable.

Fig. C-1: Aerial cable used for outside distribution.

cable, coaxial: A cable used for high frequency, consisting of two cylindrical conductors with a common axis separated by a dielectric; normally the outer conductor is operated at ground potential for shielding.

cable, control: Used to supply voltage (usually ON or OFF) to motor controls and other controllers that operate electrical circuits or apparatus. Often used in cable trays.

cable cutters: Designed to cut aluminum and copper cables but not designed to cut steel cables. See Fig. B-14.

cable, duplex: A twisted pair of cables. See Fig. C-2.

cable, festoon: A cable draped in accordion fashion from sliding or rolling hangers, usually used to feed moving equipment such as bridge cranes.

cable, hand: A mining cable used to connect equipment to a reel truck.

cable hoist: Sometimes called a coffin hoist, it is capable of lifting, lowering, stretching, and tightening.

cable joint: Where two or more lengths of cable are connected; it is sometimes referred to as a splice.

cable, parkway: Designed for direct burial with heavy mechanical protection of jute, lead, and steel wires.

cable, portable: Used to transmit power to mobile equipment.

cable, power: Used to supply current (power).

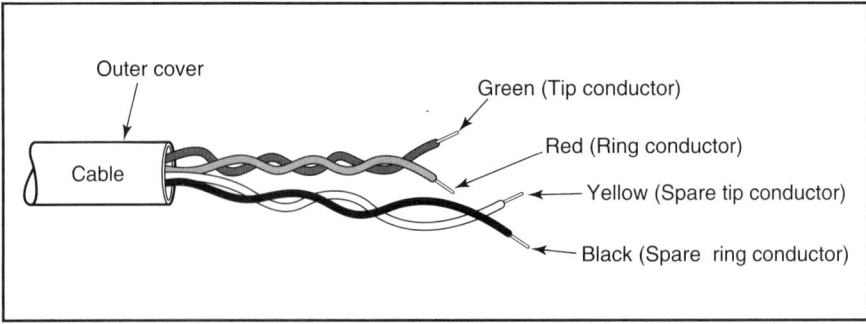

Fig. C-2: Cable, duplex. A twisted pair of cables used in telephone installation.

cable, pressure: A cable having a pressurized fluid (gas or oil) as part of the insulation: paper-fluid.

cable, ribbon: A flat multiconductor cable.

cable, service drop: The cable from the utility line to the customer's property. See Fig. C-3.

cable sheath: The protective covering that surrounds the cable assembly.

cable, signal: Used to transmit data.

cable, spacer: An aerial distribution cable made of covered conductors held by insulated spacers; designed for wooded areas.

cable, spread room: A room adjacent to a control room to facilitate routing of cables in trays away from the control panels.

cable staples: Staples specifically designed for securing electrical cable to wood beams, studs, joists, etc.

cable, submarine: Designed for crossing under navigable bodies of water; having heavy mechanical protection against anchors, floating debris, and moisture.

cable, tray: A multiconductor having a nonmetallic jacket, designed for use in cable trays; (not to be confused with type TC cable for which the jacket must also be flame retardant).

Fig. C-3: Typical overhead electric service drop supplying a residence.

Fig. C-4: Practical application of a cable tray system.

cable tray: A rigid structure to support cables; a type of raceway; normally having the appearance of a ladder and open at the top to facilitate changes. See Fig. C-4.

cable, triplexed d: Helical assembly of three insulated conductors and sometimes a bar grounding conductor.

cable, unit: A cable having pairs of cables stranded into groups (units) of a given quantity; then these groups form the core.

cable, vertical riser: Cables utilized in circuits of considerable elevation change; usually incorporate additional components for tensile strength.

cabling: Helically wrapping together of two or more insulated wires.

cad cell: A photoconductive device that has high resistance in darkness and low resistance in daylight. Used to sense light when the speed of operation is important.

cadweld: An exothermic type of weld. Also called thermoweld.

caisson: Sunken panel in a ceiling, contributing to a pattern.

calender: A machine that mixes and makes slabs of polymers by squeezing heated, viscous material between two counter-rotating rollers.

calibrate: To compare or regulate a device utilizing a recognized standard of comparison.

calorie: Heat required to raise temperature of 1 gram of water 1 degree centigrade.

calorimeter: 1) An instrument for measuring the heat generated by an electrical current in a conductor. 2) An apparatus for measuring the quantity of heat generated by friction, combustion, or other chemical change.

cambric: A fine weave of linen or cotton cloth used as insulation.

candela (cd): The basic SI unit for luminous intensity: the candela is defined as the luminous intensity of $1/600,000$ of a square meter of a blackbody at the temperature of freezing platinum.

candelabra base: Small screw-in base for decorative lamps.

candlefoot: A unit of illumination. The light given by a British standard candle at a distance of one foot. See *footcandle*.

candle power: The illuminating power of a standard sperm candle used as a measure for other illuminants.

canopy: 1) The part of a ceiling fixture that covers the wiring, junction box, and ceiling opening. 2) A light box suspended from the ceiling.

canopy switch: A small switch usually located in the canopy of an electric fixture.

cantilever: Projecting beam or member supported at only one end.

cant strip: Beveled strip placed in the angle between the roof and an abutting wall to avoid a sharp bend in the roofing material; strip placed under the lowest row of tiles on a roof to give it the same slope as the rows above it.

capacitance: The storage of electricity in a capacitor; the opposition to voltage change; the unit of measurement is the farad (f) or microfarad (mf).

capacitive phase shift: A change in the phase relationship between two quantities of current or voltage caused by insertion of a capacitor in the circuit.

capacitive reactance: The measure of resistance to the flow of an alternating current through a resistor.

capacitor: An apparatus consisting of two conducting surfaces separated by an insulating material. It stores energy, blocks the flow of direct current, and permits the flow of alternating current to a degree depending on the capacitance and frequency.

capacitor start: Refers to the use of a capacitor in series with the start winding of a motor to increase starting torque.

capacitor-start motor: A capacitor is used to give the motor a high starting torque. A centrifugal switch is used to remove the capacitor from the start winding when the motor has reached about 75% of full speed. See Fig. C-5.

capillary action: The traveling of liquid along a small interstice due to surface tension.

capstan: A rotating drum used to pull cables or ropes by friction; the cables are wrapped around the drum.

Fig. C-5: Capacitor-start motor winding schematic.

carbon black: A black pigment produced by the incomplete burning of natural gas or oil; used as a filler.

carbon dioxide: Compound of carbon and oxygen that is sometimes used as a refrigerant.

carbon film resistors: Fixed resistors that are made by coating ceramic rods with a film of carbon.

carbonizing: The reduction of a substance to carbon by subjecting it to intense heat in a closed vessel.

carpet strip: A strip attached to the floor beneath a door.

carriage: The timber or steel joist that supports the steps of a wooden stair.

carrying capacity: The greatest amount of electrical current that a conductor can safely carry, expressed in amperes. The various size wires, with their carrying capacities, are arranged in tables in the National Electrical Code. See *current-carrying capacity.*

cartridge fuse: A fuse enclosed in an insulating tube to confine the arc when the fuse blows.

cascade: The output of one device connected to the input of another.

casing: The framework about a window or door.

catepuller: Two endless belts that squeeze and pull a cable by friction.

cathode: 1) The negative electrode through which current leaves a nonmetallic conductor, such as an electrolytic cell. 2) The positive pole of a storage battery. 3) Vacuum tube—the electrode that emits electrons.

cathode rays: Streams of electrons emitted from the filament (called the cathode) of a vacuum tube under the influence of high voltage and which, by suitable means, can be brought outside the tube.

cathode-ray tube: The electronic tube that has a screen upon which a beam of electrons from the cathode can be made to create images; for example, the television picture tube.

cathodic protection: Reduction or prevention of corrosion by making the metal to be protected the cathode in a direct-current circuit.

cation: The element or positive ion that appears at the cathode or negative terminal in an electrolytic cell.

CATV: Community Antenna Television. A system of using a centralized antenna and/or satellite downlink site to receive broadcast television signals and transmit them to homes via fiber optic and or coaxial cable.

caulking: Making a joint or seam watertight or steam-tight by filling it in with rust cement or by closing the joint by means of a caulking tool. Oakum is frequently caulked into the seams of wooden vessels.

cavity wall: Wall built of solid masonry units arranged to provide airspace within the wall.

CB: Pronounced "see bee," an expression used to refer to "circuit breaker," taken from the initial letters C and B.

CC: A marking indicating that wire connectors and/or solder lugs are applicable for use with copper-clad aluminum conductors.

C-C: Center to center.

CC/CU: A marking indicating that wire connectors and/or solder lugs are applicable for use with copper-clad aluminum and copper wire.

CCA (Customer Complaint Analysis): A formal investigation of a cable defect or failure.

CEE (International Commission on Rules for the Approval of Electrical Equipment): Controls the standards for electrical products for sale in Europe; analogous to UL in USA.

cell: A single element of an electric battery, either primary or secondary.

celsius: A system in which the temperature of water's boiling point is 100 degrees and the freezing point of water is 0 degrees.

CEMF: Counter-ElectroMotive Force. The voltage produced by self-inductance.

center tap: When the connection is taken from the center of a transformer coil. See Fig. C-6.

center tapped delta: A transformer or motor circuit whereby the neutral is derived from the center tap of the phase winding. See Fig. C-6.

Centigrade scale: Temperature scale used in metric system. Freezing point of water is 0 degrees; boiling point is 100 degrees.

centimeter: One hundredth of a meter or .3937 inches.

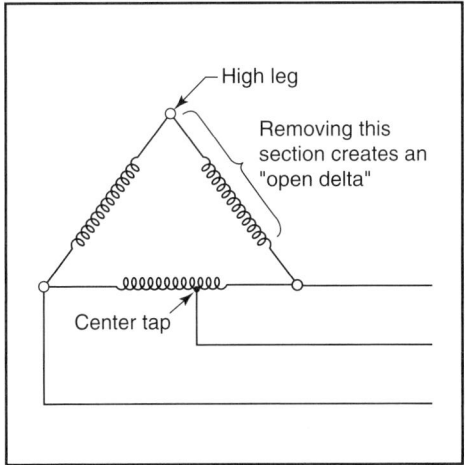

Fig. C-6: Center tap delta. The neutral is derived from the center of a transformer coil.

central processing unit (CPU): A computer circuit system that has the ability to input and output data at a very high rate of speed. The elements of the system consist of signal processing and memory elements.

centrifugal switch: A switch used on the single-phase, split-phase motor to open the starting winding after the motor has almost reached synchronous speed.

CFM: Cubic Feet per Minute.

CFR (Code of Federal Regulations): The general and permanent rules published in the Federal Register by the executive departments and agencies of the Federal Government. The Code is divided into 50 Titles that represent broad areas; Titles are divided into Chapters that usually bear the name of the issuing agency, e.g., Title 30—Mineral Resources, Chapter I = *MESA*; Title 29—Labor, Chapter XVII—*OSHA*; Title 10—Energy, Chapter I = *NRC*.

CFR Title 10, Chapter I (nuclear power): The regulations of the Federal Nuclear Regulatory Commission: a) Standards for protection against radiation. b) Licensing procedures of production and utilization facilities. c) Operators' licenses. d) Special nuclear materials. e) Nuclear material packaging for transport. f) Reactor site criteria.

cgs: A system of units: centimeter, gram, second.

chalkline: A device that holds colored powdered chalk and has a retractable string line. It is used to snap straight lines on floors, ceilings, and walls.

chamfer: A beveled edge or cut-off corner.

channel: The support base of a fluorescent fixture that holds the starter, ballast, and lamp base.

charge: The quantity of positive or negative ions in or on an object; the quantity of electricity residing on an electrostatically charged body. Unit: coulomb.

charge time: The amount of time it takes to charge a capacitor.

charged cell: A storage cell that has had direct current passed through it until the positive plate has changed chemically from $PbSO_4$ to $PbSO_2$ and the negative plate has changed from $PbSO_4$ to Pb.

charging current: Direct current applied to a storage battery to produce chemical action to charge the battery. Its direction is always the reverse of the discharge current.

charging rate: The rate of flow, in amperes, of electric current flowing through a storage battery while it is being charged.

chase: Recess in inner face of masonry wall providing space for pipes and/or ducts.

chassis: The framework to which a motor or electronic equipment is attached.

check valve: A device that permits fluid flow only in one direction.

chemical resistance test: Checking performance of materials immersed in different chemicals; loss of strength and dimensional change are measured.

chimney effect: Tendency of air or gas to rise when heated.

choke: A current filter used in a dc power supply or an inductor that has an impedance to ac current.

choke ballast: A ballast that has no transformer and is used in small fluorescent fixtures.

choke coil: A coil used to limit the flow of alternating current while permitting direct current to pass.

circline fluorescent fixture: A circular lamp that provides a considerable amount of uniform light in areas where it is desired. See Fig. C-7.

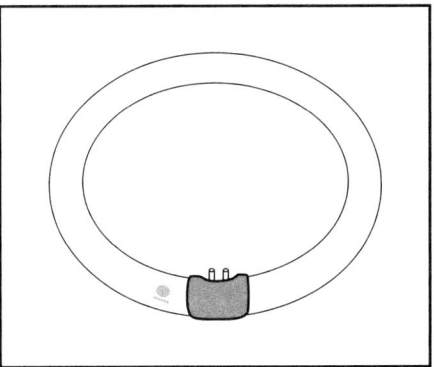

Fig. C-7: Circline fluorescent fixture is used in areas where concentrated light is important.

circuit: A closed path through which current flows from a power source through various components and back to the source. See Fig. C-8.

circuit analysis method: A technique that uses a rational sequence of trouble-shooting techniques when replacing SSR's.

circuit breaker: A resettable fuse-like device designed to protect a circuit against overloading. See Fig. C-9.

circuit foot: One foot of circuit; i.e., if one has a 3-conductor circuit, then each lineal foot of circuit would have 3 circuit feet.

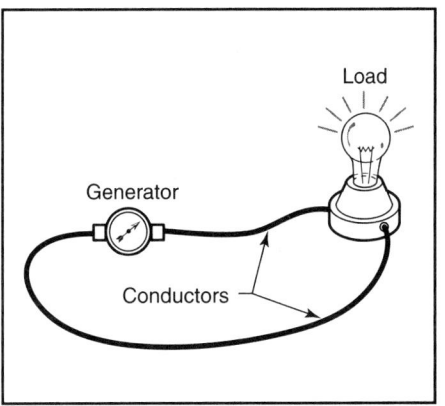

Fig. C-8: Circuit (simplified). A closed path from the generator to the load and back to the generator.

circuit reduction: A method of circuit analysis whereby a complex resistive circuit is replaced by a single resistance that is equivalent to the complex circuit.

circuit tester: A testing device that checks for correct wiring. It identifies common circuit faults and is used commonly on 120 volt 60 cycle circuits.

circular mil: The non-SI unit for measuring the cross-sectional area of a conductor.

circumstance: The environment in which a cable is located.

CL: Center Line; sometimes written C_L.

cladding: The optical coating that covers the core of the fiber and traps the light in the core.

clamp-on-meter: An ammeter used to measure current without physically interrupting the circuit. See Fig. A-6.

Fig. C-9: Operating characteristics of a typical one-pole circuit breaker.

Class 1E (nuclear power): The safety electrical systems that are essential to emergency reactor shutdown, cooling, and containment.

Class 2 (nuclear power): Items important to reactor operation but not essential to safe shutdown or isolation.

clearance: The vertical space between a cable and its conduit.

clearing time: The time from sensing an overcurrent to circuit interruption.

cleats: Small blocks of wood to which larger pieces of wood are nailed or screwed for support.

clock timer: A device with a time-delay that uses a clock to set the delay period.

closed circuit: A circuit that has a complete path and is uninterrupted, forming a complete loop for current flow.

closed loop: A description of a closed circuit when conducting circuit calculations. The algebraic sum of the voltages around any closed loop must equal zero.

closed-loop circuits: In security system circuits where the circuits are closed-looped and connected in series, if any part of that circuit is broken, a signal will be sent out.

closing die: A die used to position the individual conductors during cabling.

CO/ALR: A marking on a switch or receptacle that allows the use of copper, copper-clad, or aluminum conductors to be connected to them.

coated wire: Wire given a thin coating of another metal such as tin, lead, nickel, etc.; coating by dipping or planting; coating for protection, or to improve its properties.

coaxial cable: A cable consisting of two conductors concentric with and insulated from each other.

code: Any set of rules devised for the purpose of securing uniformity in work and for the maintaining of proper standards is usually called a code; that is, National Electrical Code, etc.

code installation: An installation that conforms to the local code and/or the national code for safe and efficient installation.

coefficient of expansion: The change in dimension due to change in temperature.

coefficient of friction: The ratio of the tangential force needed to start or maintain uniform relative motion between two contacting surfaces to the perpendicular force holding them in contact.

coil: A wire or cable wound in a series of closed loops; successive turns of insulated wire that create a magnetic field when an electric current passes through them.

coil current: Phase current through a transformer coil.

coil voltage: Phase voltage from a transformer coil.

cold bend: A test to determine cable or wire characteristics at low temperatures.

cold cathode: A cathode that does not depend on heat for electron emission.

cold chisel: 1) A metal-cutting chisel designed to penetrate any type of metal softer than itself. 2) A concrete-cutting chisel used to chip out holes in concrete.

cold joint: Improper solder connection due to insufficient heat.

cold welding: Solid-phase welding using pressure without heat.

cold work: Permanent strain produced by an external force (such as wire drawing) in a metal below its recrystallization temperature.

collapse (magnetic field): This occurs when the magnetic force drops to zero.

collar: A ceiling fixture's finished locknut.

collar beam: A beam above the lower ends of the rafters and attached to them.

collector: A transistor's semiconductor region where the base has to be connected to the same polarity.

color code: Identifying conductors by the use of color.

color temperature: Refers to the "whiteness" of the incandescent light.

combination circuit: A circuit that consists of a combination of series and parallel branches.

combination square: A square designed to have a moveable blade marked in metrics or inches. The square has an apparatus to lock the blade in position.

combustible: Having the capability to catch fire fairly easily.

combustion: The chemical union of a combustible substance with oxygen, resulting in the production of heat.

come along: A cable grip (usually of tubular basketweave construction that tightens its grip on the cable as it is pulled) with a pulling "eye" on one end for attaching to a pull-rope for pulling conductors into conduit or other raceway.

comeback bend: A bend made by conduit benders. "Comeback" one half the gain to mark center of the bend and to place benders in right location for bending.

comfort zone: Area on psychrometric chart that shows conditions of temperature, humidity, and sometimes air movement in which most people are comfortable.

common: An identified terminal marked to receive the hot lead on a three-way switch.

common failure mode (nuclear power): An event causing redundant equipment to be inoperable.

commutating pole: An electromagnetic bar inserted between the pole pieces of a generator to offset the cross magnetization of the armature currents.

commutator: Device used on electric motors or generators to maintain a unidirectional current.

comparator: An operational amplifier used as a level detector. Detects when one voltage becomes greater than another.

Compax® die: A wire drawing die made by *GE* of sintered diamond.

compensated wattmeter: A wattmeter that has a compensating coil connected in series with the potential coil for the purpose of correcting the error caused by the absorption of power due to mechanical operation of the potential coil.

compensating coil: A coil that serves to compensate for the mechanical friction in the moving coil of a meter.

completed circuit: Also called closed circuit. One that has been made or closed.

compound: A combination of two or more parts or elements.

compound fill: An insulation that is poured into place while hot.

compound generator: A generator that has both series and shunt fields within itself.

compound wound: A generator or motor having a part of a series-field winding wound on top of a part of a shunt-field winding on each of the main pole pieces.

compressibility: A density ratio determined under finite testing conditions.

compression lug or splice: Installed by compressing the connector onto the strand, hopefully into a cold weld.

compressor: The pump of a refrigerating mechanism that draws a vacuum on the cooling side of a refrigerant cycle and squeezes or compresses the gas into the condensing side of the cycle.

computed load: The calculated electrical load of a building or circuit based on National Electrical Code requirements.

computer: An electronic apparatus: 1) For rapidly solving complex and involved problems, usually mathematical or logical. 2) For storing large amounts of data.

concealed: Rendered inaccessible by the structure or finish of the building. Wires in concealed raceways are considered concealed, even though they may become accessible by withdrawing them.

concentric bends: The radius of each bend of conduit must be increased to keep a uniform look when two or more conduits run parallel with each other and have a common center.

concentricity: The measurement of the center of the conductor with respect to the center of the insulation.

condenser: The accumulator of electrical energy, as in a capacitor. A vessel in which the condensation of gases is effected.

conductance: The ability of material to carry an electric current; that is, the ease with which a conductor carries an electric current. The opposite of resistance.

conduction: The flow of an electric current through a conducting body, such as a copper wire.

conductive compensation: An AC universal motor that has a winding in series with the armature to improve its efficiency.

conductivity: The relative ability of materials to carry an electrical current.

conductor: Any substance that allows energy flow through it with the transfer being made by physical contact but excluding net mass flow.

conductor ampacity: A conductor's current carrying capacity.

conductor, bare: Having no covering or insulation whatsoever.

conductor, covered: A conductor having one or more layers of nonconducting materials that are not recognized as insulation under the National Electric Code.

conductor, insulated: A conductor covered with material recognized as insulation.

conductor load: The mechanical loads on an aerial conductor—wind, weight, ice, etc.

conductor, plain: Consisting of only one metal.

conductor, segmental: Having sections isolated, one from the other and connected in parallel; used to reduce ac resistance.

conductor, solid: A single wire.

conductor, stranded: Assembly of several wires, usually twisted or braided.

conductor stress control: The conducting layer applied to make the conductor a smooth surface in intimate contact with the insulation; formerly called extruded strand shield (ESS).

conduit: A tubular raceway such as electrical metallic tubing (EMT); rigid metal conduit, rigid nonmetallic conduit, etc. See Fig. C-10.

conduit body: A device or fitting used with conduit that allows turning of corners and accessibility to the conductors.

conduit fill: The number of conductors allowed by the National Electrical Code to be installed in a conduit that will safely carry rated amps without overheating.

conduit fittings: The connecting devices that complete a conduit system.

conduit, rigid metal: Conduit made of Schedule 40 pipe, normally 10-foot lengths.

conduit run: A complete electrical raceway that runs between two electrical devices.

Condulet: A trade name for conduit fitting.

configuration, cradled: The geometric pattern that cables will take in a conduit when the cables are pulled in parallel and the ratio of the conduit ID to the 1/C cable OD is greater than 3.0.

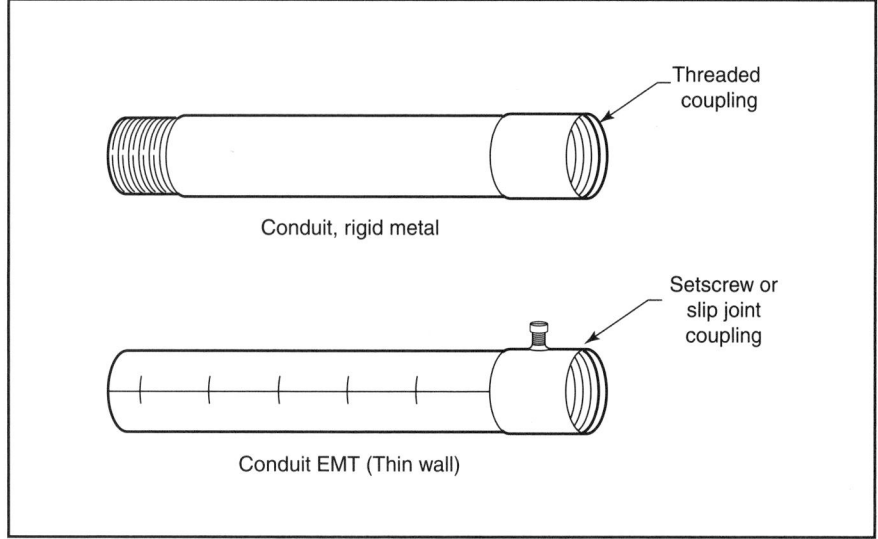

Fig. C-10: Conduits made of rigid metal and EMT.

configuration, triangular: The geometric pattern that cables will take in a conduit when the cables are triplexed or are pulled in parallel with the ratio of the conduit ID to the 1/C cable OD less than 2.5.

connected load: The total watts needed to supply a building or circuit if all of the loads in the building or circuit are drawing power at once.

connection: That part of a circuit that has negligible impedance and joins components or devices.

connection, delta: Interconnection of three electrical-equipment windings in delta (triangular) fashion.

connection (nuclear power): A cable terminal, splice, or seal at the interface of the cable and equipment.

connection, star: Interconnection of three electrical-equipment windings in star (wye) fashion.

connector, pressure: A connector applied using pressure to form a cold weld between the conductor and the connector.

connector, reducing: Used to join two different size conductors.

Fig. C-11: Several types of wire connectors.

connectors: 1) Devices used to physically and electrically connect two or more conductors. See Fig. C-11. 2) Special connectors that join fiber optics to each other, protect the fiber optic material, and do not allow light to escape. 3) Join conduit to boxes, panels, and other electrical equipment.

constant current: A type of power system in which the same amount of current flows through each utilization equipment, used for simplicity in street lighting circuits.

constant voltage: The common type of power in which all loads are connected in parallel with different amounts of current flowing through each load.

contact: A device designed for repetitive connections.

contactor: A type of relay.

containment (nuclear power): The safety barrier designed to prevent the release of radioactive material in case of reactor accident.

continuity: The state of being whole, unbroken.

continuity tester: A tester that is designed to check the electrical system or circuit to ascertain that there is a complete path.

continuous beam: A beam that rests on more than two supports.

continuous duty: That demand on a system requiring operation at a constant load for an indefinite period of time.

continuous load: 1) NEC—in operation three hours or more. 2) Nuclear power—8,760 hours/year (scheduled maintenance outages permitted).

continuous vulcanization (CV): A system utilizing heat and pressure to vulcanize insulation after extrusion onto wire or cable; the curing tube may be in a horizontal or vertical pole.

contract modifications: Change orders, additional instructions, and/or field orders that are issued after the bidding contract has been signed. Generally these are issued by the architect or owner.

control: Automatic or manual device used to stop, start, and/or regulate flow of gas, liquid, and/or electricity.

controller: A device or group of devices that serves to govern in some predetermined manner the electric power delivered to the apparatus to which it is connected.

controller function: The controller selected depends on the type of application. The controller monitors the inputs of devices and activates the outputs of other devices depending on the needs of the application.

controller service: A controller that has a particular function.

control transformer: A supply voltage goes to the main system and a step-down transformer is used to supply voltage to the control system.

control, temperature: A thermostatic device that automatically stops and starts a motor, the operation of which is based on temperature changes.

convection: The transfer of heat to a fluid by conduction as the fluid moves past the heat source.

convenience outlet: A point on the wiring system at which current is taken to supply portable 120-volt appliances such as tv sets, toasters, etc.

converter: A device that changes the standard 60Hz AC power to any other frequency or AC voltage to DC voltage.

cook annealer: An annealer using heavy electrical current through the conductor as the heat source.

cook buncher: A buncher using controlled diameter and wire position.

cooling tower: Device that cools water by water evaporation in air. Water is cooled to the wet bulb temperature of air.

coordination: The selection of system components to prevent the failure of the whole system due to cascading; limiting system failure by activation of the fewest overcurrent devices, hopefully to one.

co-polymer: A polymer having two "repeating units."

copper: A word used by itself to refer to copper conductors. Examples: "A circuit of 500 MCM (kcmil) copper" or "the copper cost of the circuit." It is a good conductor of electricity, easily formed, easily connected to itself and other metals used in electrical wiring.

Copper Development Association: A manufacturer's group to promote the use of copper.

copper, electrolytic: Copper of high purity, refined by electrolysis, used for electrical conductors.

copper loss: The energy dissipated in the copper conductors of a circuit, due to heat loss of I^2R produced by current flow through the conductor. Term is sometimes used to refer to the same type of loss in aluminum circuit conductors.

Copperweld®: This is the trade name for a conductor composed of a steel core with a heavy copper coating. It is used where strength is needed such as large, long, vertical risers; most commonly used for ground rods.

cord: A small flexible conductor assembly, usually jacketed.

cord-connected unit: Electrical units that are plugged into a receptacle.

cord set: A cord having a wiring connector on one or more ends.

core: The portion of a foundry mold that shapes the interior of a hollow casting.

core (cable): 1) The portion of an insulated cable under a protective covering. 2) The optical fiber that light travels through. It is the center of the fiber cable.

cored hole: At the time of casting, a sand core is placed in the mold so that the metal flows around it. When the casting is cold, the sand core is broken away, leaving the hole.

core disks: Thin circular disks of sheet iron used for building laminated armature cores.

core loss: The electric loss occurring in the core of an armature or transformer due to eddy currents, hysteresis, and like influences.

corona: A low energy electrical discharge caused by ionization of a gas by an electric field.

correction factors: The correction factors that affect the current carrying capacity of a conductor when the ambient temperature rises above 30°C. Table 310-16 of the NEC®.

corrective maintenance: Defining why a part or a piece of equipment failed and correcting the problem. This is *not* parts changing.

corrosion: The deterioration of a substance (usually a metal) because of a reaction with its environment.

cosine: The cosine is the adjacent side divided by the hypotenuse when finding the degree of an angle.

coulomb: A unit of electrical quantity. The derived SI unit for quantity of electricity or electrical charge: one coulomb equals one ampere-second.

counterbore: A tool that enlarges an already-machined round hole to a certain depth. The pilot of the tool fits in the smaller hole, and the larger part counterbores or makes the end of the hole larger.

counter emf: The voltage opposing the applied voltage and the current in a coil; caused by a flow of current in the coil; also known as back emf; induced voltage.

coupled-inductance: Voltage induced in one circuit by current changes in a second circuit.

coupling: The means by which signals are transferred from one circuit to another. The mechanical connector between two pieces of conduit.

coupon: A piece of metal for testing, of specified size; a piece of metal from which a test specimen may be prepared.

cove: A cove-shaped baffle connected to a valance that directs light to a specific area.

CPE (Chlorinated PolyethylenE): A plastic for cable jackets.

cramp: Iron rod with ends bent to a right angle; used to hold blocks of stone together.

crawl space: Shallow space between the first tier of beams and the ground (no basement).

crazing: Fine cracks that may extend in a network on or under the surface of a material; usually occurs in the presence of an organic liquid or vapor.

crescent wrench: See *adjustable wrench.*

crimp connector: A metal device designed to be compatible with the conductors that are being connected to each other. It uses a special crimping tool to make a permanent connection.

cross head: The mechanism on an extruder where the material is applied; it holds the die, guider, and core tube; usually just called "head."

cross-link: To cure by linking molecules together in a polymer—either by using chemical cross-linking agents or radiation.

cross-linked polyethylene: Thermosetting polyethylene that has better physical properties than plain polyethylene: used as an insulation having good physical properties.

cross talk: Undesired pickup of signals by a second circuit.

CRT: Cathode Ray Tube.

crystal: A solid composed of atoms, ions, or molecules arranged in a pattern that is periodic in three dimensions.

CSA: 1) Cross Sectional Area. 2) Canadian Standards Association.

CT: Pronounced "see tee"; refers to current transformer, taken from the initial letters C and T.

CU: Copper.

cure: To change the properties of a polymeric system into a more stable, usable condition by the use of heat, radiation, or reaction with chemical additives.

current (I): The time rate of flow of electric charges; unit: ampere.

current-carrying capacity: The current in amperes a conductor can carry continuously under the conditions of use without exceeding its temperature rating.

current, charging: The current needed to bring the cable up to voltage; determined by capacitance of the cable; after withdrawal of voltage, the charging current returns to the circuit; the charging current will be 90° out of phase with the voltage.

current density: Current per unit of a cross-sectional area.

current-induced: Current in a conductor due to the application of a time-varying electromagnetic field.

current, leakage: That small amount of current that flows through insulation whenever a voltage is present and heats the insulation because of the insulation's resistance; the leakage current is in phase with the voltage and is a power loss.

current limiting: A characteristic of short-circuit protective devices, such as fuses, by which the device operates so fast on high short circuit currents that less than a quarter wave of the alternating cycle is permitted to flow before the circuit is opened, thereby limiting the thermal and magnetic energy to a certain maximum value, regardless of the current available.

current rating: The maximum amount of amps a device can safely and properly work without doing damage to the device.

current sinking output: An output device that uses NPN transistors to control a switching procedure.

current sourcing output: A device that senses a target by using a PNP transistor.

current transformer: A current transformer is used when the range of an AC ammeter needs to be extended.

curtain wall: A thin wall supported independently of the wall below, by the structural steel or concrete frame of the building.

cut-in: The connection of electrical service to a building, from the power company line to the service equipment, e.g., "the building was cut-in" or "the power company cut-in the service."

cut-in-card: The certificate of approval issued by the electrical inspection authority to the electrical contractor, to be given to the power company as evidence that the building electrical system is safe for connection or "cut-in" by the power company.

cut nails: Machine-cut iron nails as distinguished from wire nails.

cutoff region: A transistor has a predetermined point that causes it to switch off and stop current flow.

cutout: A fuse holder that may be used to isolate part of a circuit.

cutout box: A surface mounting enclosure with a cover equipped with swinging doors, used to enclose fuses.

cutover: Changing from one reel to another without stopping the manufacturing process.

cut resistance: The ability of a material to withstand mechanical pressure without rupture or becoming ineffective.

cycle: 1) An interval of space or time in which one set of events or phenomena is completed. 2) A set of operations that are repeated regularly in the same sequence. 3) When a system in a given state goes through a number of different processes and finally returns to its initial state. 4) The part of a periodic wave form that does not repeat itself. One complete cycle.

cyclic aging: A test on a closed loop of cable having voltage applied, and induced current applied in cycles to cause expansion and contraction; simulates cable operating in a dry environment.

D

damper: Valve for controlling air flow.

damping: The dissipation of energy with time or distance.

damping coil: A coil mounted near a galvanometer to produce a damping effect; i.e., to bring the needle quickly to a point of rest after deflection.

damp location: A location that is subject to dampness. The electrical components in this area must be marked suitable for damp locations.

dampness control: A device that senses dampness and often controls both humidifiers and/or dehumidifiers. Often referred to as a humidistat.

Daniell cell: A closed-circuit type of primary cell.

D'Arsonval galvanometer: A very sensitive periodic or deadbeat galvanometer in which the indicating coil is suspended in the field of a powerful horseshoe magnet.

dashpot timer: A device that is able to control, because of an orifice, the speed at which air or liquid either leaves or enters a container.

data I/O: Programmable controllers (PLCs) that have input and output functions.

DBE (Design Basics Event) (nuclear power): Postulated abnormal events used to establish the performance requirements of the structures, systems, and components.

dead: 1) Not having electrical charge. 2) Not having voltage applied.

dead beat: Instruments in which indicators come promptly to a position of rest due to heavy damping.

dead-end: A mechanical terminating device on a building or pole to provide support at the end of an overhead electric circuit. A dead-end is also the term used to refer to the last pole in the pole line. The pole at which the electric circuiting is brought down the pole to go underground or to the building served.

dead-front: The removable portion of the front of an electrical panel that prevents contact with energized parts which, when removed, allows the electrician to service the equipment.

dead load: A load whose pressure is steady and constant.

deadman: Reinforced concrete anchor set in earth and tied to the retaining wall for stability.

deadman's switch: A switch necessitating a positive action by the operator to keep the system or equipment running or energized.

debug: To examine or test a procedure, routine, or equipment for the purpose of detecting and correcting errors, especially during start-up.

decay (nuclear power): The transmutation of a nucleus to a stable energy condition.

decibel: A unit for measuring sound intensity, named in honor of Alexander Graham Bell. When sound or noise is created energy is given off which is measured in decibels; i.e., the noise of an airplane engine measures 120 decibels.

decorative lighting: Lighting generally used outdoors to create or enhance a mood or space.

deeping: Cutting out to a depth; placing comparatively far below the surrounding surface.

defeater: A means to deactivate a safety interlock system.

defense in depth (nuclear power): A basic design philosophy to keep nuclear power plants safe during normal operations and the worst imagined accidents. There are three levels of defense: 1) Accident prevention, quality assurance, redundancy, inspection testing. 2) Protection devices and systems. 3) Safety systems to function in case 1 and 2 fail.

definite time: See *dashpot timer*.

deflection: Deviation of the central axis of a structural beam from normal when the beam is loaded.

deflection plate: The part of a certain type of electron tube that provides an electrical field to produce deflection of an electron beam.

degree indicator: (pipe bending) A tool designed to show the exact degree of a bend when the bend is being constructed.

delta connection: The connection of the circuits in a three-phase system in which the terminal connections are triangular like the Greek letter delta. See Fig. D-1.

demagnetization: The process of removing magnetism from a magnetized substance.

demagnetized: When the magnetic field of a material has been removed. This can be accomplished by heating or striking.

demand: 1) The measure of the maximum load of a utility's customer over a short period of time. 2) The load integrated over a specified time interval.

demand factor: For an electrical system or feeder circuit, this is a ratio of the amount of connected load (in kVA or amperes) that

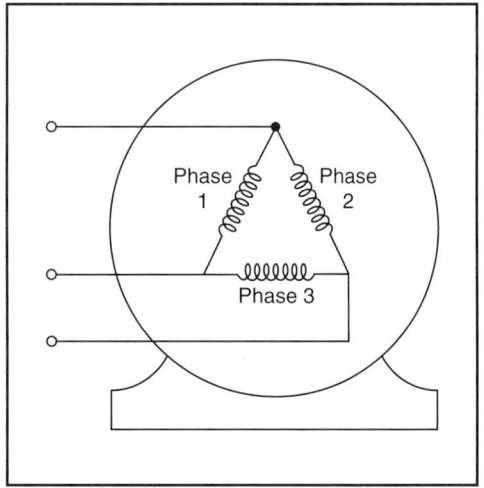

Fig. D-1: Delta connection of a delta-wound motor.

will be operating at the same time to the total amount of connected load on the circuit. An 80% demand factor, for instance, indicates that only 80% of the connected load on a circuit will ever be operating at the same time. Conductor capacity can be based on that amount of load.

demand load: The probable load that is connected to an electrical system or part of a system.

demarcation point: The location where the local telephone company ends its responsibility for maintenance and the homeowner or business accepts its responsibility for maintenance. See Fig. D-2.

Fig. D-2: Demarcation point. The point at which the local telephone company ends its responsibility and the customers begin their responsibility.

demonstration (nuclear power): A course of reasoning showing that a certain result is a consequence of assumed premises; an explanation or illustration.

density: Closeness of texture or consistency.

depolarization: The process of preserving the activity of a voltaic cell by preventing polarization.

depolarizer: An oxidizing substance used for fixing the hydrogen derived from the decomposition of the acid by the zinc in primary cells.

derating: The intentional reduction of stress/strength ratio in the application of a material; usually for the purpose of reducing the occurrence of a stress-related failure.

derating factor: A factor used to reduce ampacity when the cable is used in environments other than the standard, or when more than the standard number of conductors is installed in a raceway.

details (drawings): Drawings that have been enlarged to show specific areas or aspects of a remodel and/or installation.

detection: The process of obtaining the separation of the modulation component from the received signal.

developed length: The absolute length of a conduit or pipe that is going to be bent.

deviation: Departure from the exact or from a set standard.

device: An item intended to carry or help carry, but not utilize, electrical energy. See Fig. D-3.

dew point: The temperature at which vapor starts to condense (liquify) from a gas-vapor mixture at constant pressure.

DIAC: A bidirectional diode used primarily as a triggering device.

diagonal pliers: Pliers designed to cut cables and conductors in a limited space. Some can be purchased with a stripping slot for stripping insulation from conductors.

diagram (electrical): A working drawing of an object that is to be constructed.

Dielectric constant 63

Fig. D-3: The single-pole switch and duplex receptacle are known as wiring devices since they do not actually use electricity for operation.

dial: A graduated plate, usually circular or oval, on which a reading is indicated by a needle or pointer.

die: 1) Wire: a metal device having a conical hole that is used to reduce the diameter of wire drawn through the die or series of dies. 2) Extruder: the fixed part of the mold. 3) An internal screw used for cutting an outside thread.

dielectric: An insulator or a term referring to the insulation between the plates of a capacitor.

dielectric absorption: The storage of charges within an insulation; evidenced by the decrease of current flow after the application of dc voltage.

dielectric breakdown (capacitor): Failure of an insulator to prevent current flow from one plate to another.

dielectric constant: The ratio of the conductivity of a dielectric for electrostatic lines of force to that of air.

dielectric dispersion: The change in relative capacitance due to change in frequency.

dielectric heating: The heating of an insulating material by ac induced internal losses; normally frequencies above 10 mHz are used.

dielectric loss: The time rate at which electrical energy is transformed into heat in a dielectric when it is subjected to a changing electric field.

dielectric phase angle: The phase angle between the sinusoidal ac voltage applied to a dielectric and the component of the current having the same period.

dielectric strength: The maximum voltage that an insulation can withstand without breaking down; usually expressed as a gradient—vpm (volts per mil).

differential motor: A motor with a compound-wound field, in which the series and shunt coils oppose each other.

digital logic: Solid state control devices called "gates" that use a binary form of logic. The more widely used gates are AND, NAND, OR, NOR, and INVERTER.

digital meter (DPM): A meter that always gives its reading with a digital display. See Fig. D-4.

digital signaling system: An alarm system that embodies digital alarm communicator receivers and digital alarm communicator transmitters.

dimension: A definite measurement shown on a

Fig. D-4: Digital multimeter is used for accurate electrical readings.

drawing, as length, width, or thickness. Dimensions should be given on all working drawings, whether drawn to scale or not, but not on assembly drawings, except "over-all" dimensions.

dimensioning: Indicating on a drawing the sizes of various parts.

dimension line: A line on a drawing that indicates to what part of line the dimension has reference.

dimmer control: A device used in residential and commercial units to lower the brightness of light. The two basic types are autotransformer dimmers and electronic dimmers.

diode: A device having two electrodes: the cathode and the plate or anode—used as a rectifier and detector. See Fig. D-5.

dipping (rope): A method used to keep a rope end from fraying.

direct-burial: Cabling or equipment that can have direct contact with the ground without having additional protection.

direct connected units: Electrical units that are wired directly to the branch circuit.

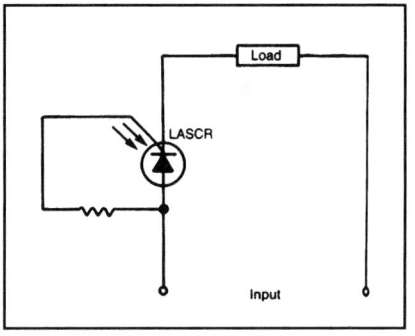

Fig. D-5: Schematic drawing of a diode.

direct current (dc): Electricity that flows only in one direction; produced by a battery and dc generators.

direct drive: A compact arrangement of driving a generator by direct connection with the prime mover or the load, avoiding the use of shafts and belts.

direction of lay: The lateral direction, designated as left-hand or right-hand, in which the elements of a cable run over the top of the cable as they recede from an observer looking along the axis of the cable.

direction of magnetic flux: The direction is always from the north to the south pole in a magnetic field.

disconnect: A switch for disconnecting an electrical circuit or load (motor, transformer, panel) from the conductors that supply power to it, e.g., "He pulled the motor disconnect," meaning he opened the disconnect switch to the motor.

disconnecting means: A device, a group of devices, or other means whereby the conductors of a circuit can be disconnected from their supply source.

discrete I/O: A programmable controller that allows an individual stand-alone machine to have its own capability to manufacture parts.

dispersion: Holding fine particles in suspension throughout a second substance.

displacement current: An expression for the effective current flow across a capacitor.

dissipation factor: Energy lost when voltage is applied across an insulation because of capacitance: the cotangent of the phase angle between voltage and current in a reactive component; because the shift is so great, we use the complement (angle) of the angle ϕ which is used for power factor; dissipation factor = tan = cot ϕ: is quite sensitive to contamination and deterioration of insulation: also known as power factor (of dielectrics).

distortion: Unfaithful reproduction of signals.

distribution box: A large metal box used in conduit installation as a center of distribution.

distribution line: An exterior supply line from which individual installations are supplied.

distribution, statistical analysis: A statistical method used to analyze data by correlating data to a theoretical curve to a) Test validity of data. b) Predict performance at conditions different from those used to produce the data: The normal distribution curve is most common.

diversity factor: The ratio of the sum of load demands to a system demand. See *demand factor*.

DOAL: Diameter Overall.

DOC: Diameter Over Conductor; note that for cables having a stress control, the diameter over the stress control layer becomes DOC.

documents (nuclear power): Pertaining to Class 1E equipment and systems.

dog-leg bend: A term given to conduit bends that are generally less than 45°.

DOI: Diameter Over Insulation.

DOIS: Diameter Over Insulation Shield.

DOJ: Diameter Over Jacket.

donkey: A motor-driven power machine on legs or wheels used for threading and/or cutting conduit. Sometimes called "mule."

DOSC: Diameter Over Stress Control.

dose, radiation: The amount of energy per unit mass of material deposited at each point of an object undergoing radiation.

double-insulated: Two types of insulation on a piece of equipment are used to guard against heat or electric shock.

double-pole switch: A device that switches two separate electrical paths at the same time.

double-strength glass: One-eighth inch thick sheet glass (glass rated as single strength is $1/10$ inch thick).

dovetail: An interlocking joint.

downconverter: The conversion of a satellite signal (high frequency) to a signal that a television set can interpret (low frequency).

down-link: The signal received from an orbiting satellite to an earth-based satellite dish.

draft indicator: An instrument used to indicate or measure chimney draft or combustion gas movement.

draft stop or fire stop: Obstructions placed in air passages to prevent the passage of flames up or across a building.

draintile: Hollow tile used for draining wet places.

drain wire: A bonding wire laid parallel to and touching shields.

drawing: Reducing wire diameter by pulling through dies.

drawing, block diagram: A simplified drawing of a system showing major items as simplified blocks; normally used to show

Fig. D-6: A power-riser (block) diagram used to show the arrangement of panels and feeders for an electrical system.

how the system operates and the relationship between major items. See Fig. D-6.

drawing, electrical: Consists of lines, symbols, dimensions, and notations to accurately convey an engineer's design to workmen who install the electrical system on the job. A means of conveying a large amount of exact, detailed information in an abbreviated language.

drawing, line schematic (diagram): Shows how a circuit works. See Fig. D-7.

drawing, plot or layout: Shows the "floor plan." See Fig. D-8.

drawing, wire diagram: Shows how electrical devices are interconnected. See Fig. D-9.

drill: A circular tool used for machining a hole.

drill motor: Portable electric drills that come in different sizes and different types. The chuck determines the size of the drill motor and ¼", ⅜", and ½" are the most common sizes. The most popular is the pistol grip, which has variable speeds and is reversible. See Fig. D-10.

drip: Projecting horizontal course sloped outward to throw water away from a building.

Fig. D-7: A schematic diagram shows how an electrical circuit works.

Fig. D-8: A floor plan of a building is drawn as if the roof was removed and the viewer is looking down at the building from above.

drip loop: An intentional sag placed in service entrance conductors where they connect to the utility service drop conductors on overhead services; the drip loop will conduct any rain water to a point lower than the service head, to prevent moisture being forced into the service conductors by hydrostatic pressure and then running through the service head into the service conduit or cable. See Fig. D-11.

Fig. D-9: A wiring diagram shows how devices are connected in an electrical circuit.

Fig. D-10: Drill motor (hand held). Common sizes are ¼-inch, ⅜-inch, and ½-inch drill motors.

dropping resistor: A resistor that is used to decrease a given amount of current by an equal amount of voltage dropped across a resistor.

drum: The part of a cable reel on which the cable is wound.

drum armature: A generator or motor armature having its coils wound longitudinally or parallel to its axis.

drum switch: A rotary design manual switch that is used for frequent stop, start, or reversing of ac or dc motors. See Fig. D-12.

dry: Not normally subjected to moisture.

dry battery: An electric battery made up of a number of dry voltaic cells. The term is often wrongly applied to a single dry cell.

dry bulb: An instrument with a sensitive element that measures ambient (moving) air temperature.

dry cell: A primary cell that does away with the liquid electrolyte so that it may be used in any position.

dry location: Electrical equipment that must be protected from permanent weather damage.

Fig. D-11: A drip loop conducts rain water to a point lower than the service head.

Fig. D-12: Drum switch. A manual switch used when frequent starts and stops are required.

dry niche: Below-water-level-housing for pool lighting in which the water level has to be lowered to change out the light bulbs or relamped from the rear. The niche must be water tight.

drywall: Interior wall construction consisting of plasterboard, wood paneling, or plywood nailed directly to the studs without application of plaster.

dual-element fuses: Fuses that come in cartridge and plug type and are constructed with two fusible elements, overload and short-circuit. The amount of excess current that flows depends on which element opens.

dual extrusion: Extruding two materials simultaneously using two extruders feeding a common cross head.

duct: A tube or channel through which air is conveyed or moved.

duct bank: Several underground conduits grouped together. See Fig. D-13.

ductility: The ability of a material to deform plastically before fracturing.

Fig. D-13: Cross-section of a duct bank utilizing grouped conduit to house conductors.

Fig. D-14: Duplex receptacle (15 amperes).

dumbbell: A die-cut specimen of uniform thickness used for testing tensile and elongation of materials.

duplex receptacle: A receptacle that is designed with two plug-in outlets. See Fig. D-14.

durometer: An instrument to measure hardness of a rubber-like material.

duty, continuous: A service requirement that demands operation at a substantially constant load for an indefinitely long time.

duty, intermittent: A service requirement that demands operation for alternate intervals of load and no load, load and rest, or load, no load, and rest.

duty, periodic: A type of intermittent duty in which the load conditions regularly reoccur.

duty, short-time: A requirement of service that demands operations at loads and for intervals of time, both of which may be subject to wide variation.

dwarf partition: Partition that ends short of the ceiling.

dwell: A planned delay in a timed control program.

dwelling: A term of rather broad use, meaning a house or residence; the *National Electrical Code®* recognizes a house as a single-family dwelling and an apartment as a multi-family dwelling. See Fig. D-15.

Fig. D-15: An apartment complex is recognized by the *National Electrical Code®* as a multi-family dwelling.

dynamic: A state in which one or more quantities exhibit appreciable change within an arbitrarily short time interval.

dynamic breaking: A method of braking generally used with dc motors. This type of braking reconnects the motor to act as a generator after the motor is turned off.

dynamo: An electrical machine for converting mechanical energy into electrical energy; an electrical generator, especially for producing direct current.

dynamometer: A device for measuring power output or power input of a mechanism.

E: Symbol for voltage, electrical pressure, electromotive force.

eaves: The projecting edges of a roof.

EC: Electrical Conductor of Aluminum.

eccentricity: 1) A measure of the entering of an item within a circular area. 2) The percentage ratio of the difference between the maximum and minimum thickness to the minimum thickness of an annular area.

ECCS (Emergency Core Cooling System) (nuclear power): A system to flood the fueled portion of the reactor and remove the residual heat produced by radioactive decay.

ECKO: Eddy Current Kill Oscillator. It works on the principle that an oscillator will function in a variable magnetic field even if a target is not present.

eddy currents: Circulating currents induced in conducting materials by varying magnetic fields; usually considered undesirable because they represent loss of energy and cause heating.

edge-emitting diode (E-LED): A semiconductor chip whose LED is emitted from the edge of the chip.

Edison base: The standard screw base used for ordinary lamps and Edison-base plug fuses.

EEI: Edison Electric Institute.

effective temperature: Overall effect on a person of air temperature, humidity, and air movement.

effective value: AC volts RMS and AC current RMS are interchangeable with effective value.

efficiency: 1) The ratio of the output to the input. 2) Output divided by input equals the efficiency ratio.

EIA: Electronics Industries Association.

elasticity: That property of recovering original size and shape after deformation.

elastomer: A material which, at room temperature, stretches under low stress to at least twice its length and snaps back to the original length upon release of stress.

elbow: A short conduit that is bent—usually at a 90-degree angle.

elbow-connector: Manufactured piece of a conduit or raceway system that forms a 90° turn. See Fig. E-1.

electrical degree: One three-hundred-sixtieth of an ac cycle.

electrical interlocking: When a device or circuit prevents another device or circuit from functioning.

electrical noise: Undesirable signals that invade electrical circuits. Power lines and I/Os are the main source of these undesirable signals.

electrical symbols: Graphical symbols used on electrical drawings. In preparing electrical drawings, most engineers and designers used symbols adopted by the

Fig. E-1: Elbow connector is used whenever a 90-degree turn is required when connecting to an electrical box.

United States of America Standards Institute (USASI). However, many designers frequently modify these standard symbols to suit their own needs. For this reason, most electrical working drawings for building construction will have a symbol list or legend. The symbols shown in Fig. E-2 (beginning on page 79) are typical of those found on electrical drawings.

electric controller: A device that is connected to a piece of equipment and controls its electric power.

electric defrosting: Use of electric resistance heating coils to melt ice and frost off evaporators during defrosting.

electric discharge tubing: Glass or plastic tubing formed into designs to be used as display lighting. The tubing is filled with various types of inert gases.

electric field: The invisible force field that is between electric charges.

electric heating: House heating system in which heat from electrical resistance units is used to heat rooms. See Fig. E-3 on page 84.

electricians knife: A knife that generally has a splitting blade that locks in the open position. It is excellent for skinning the coverings off electric cables.

electricity: Relating to the flow or presence of charged particles; a fundamental physical force or energy.

electric sign: A fixed, stationary, or portable self-contained, electrically illuminated utilization equipment with words or symbols designed to convey information or attract attention.

electric water valve: Solenoid type (electrically operated) valve used to turn water flow on and off.

electrocution: Death caused by electrical current through the heart, usually in excess of 50 ma. See Fig. E-4 on page 85.

electrode: A conductor through which current transfers to another material.

electrolysis: The production of chemical changes by the passage of current from an electrode to an electrolyte or vice versa.

electrolyte: A liquid or solid that conducts electricity by the flow of ions.

Electrical Reference Symbols

ELECTRICAL ABBREVIATIONS
(Apply only when adjacent to an electrical symbol.)

Central Switch Panel	CSP
Dimmer Control Panel	DCP
Dust Tight	DT
Emergency Switch Panel	ESP
Empty	MT
Explosion Proof	EP
Grounded	G
Night Light	NL
Pull Chain	PC
Rain Tight	RT
Recessed	R
Transfer	XFER
Transformer	XFRMR
Vapor Tight	VT
Water Tight	WT
Weather Proof	WP

ELECTRICAL SYMBOLS

Switch Outlets

Single-Pole Switch	S
Double-Pole Switch	S_2
Three-Way Switch	S_3
Four-Way Switch	S_4
Key-Operated Switch	S_K
Switch and Fusestat Holder	S_{FH}
Switch and Pilot Lamp	S_P
Fan Switch	S_F
Switch for Low-Voltage Switching System	S_L
Master Switch for Low-Voltage Switching System	S_{LM}
Switch and Single Receptacle	⊖S
Switch and Duplex Receptacle	⊜S
Door Switch	S_D
Time Switch	S_T
Momentary Contact Switch	S_{MC}
Ceiling Pull Switch	Ⓢ
"Hand-Off-Auto" Control Switch	HOA
Multi-Speed Control Switch	M
Push Button	▪

Receptacle Outlets

Where weather proof, explosion proof, or other specific types of devices are to be required, use the upper-case subscript letters. For example, weather proof single or duplex receptacles would have the uppercase WP subscript letters noted alongside of the symbol. All outlets should be grounded.

Single Receptacle Outlet	⊖
Duplex Receptacle Outlet	⊜
Triplex Receptacle Outlet	⊕
Quadruplex Receptacle Outlet	⊕

Fig. E-2: Electrical symbols.

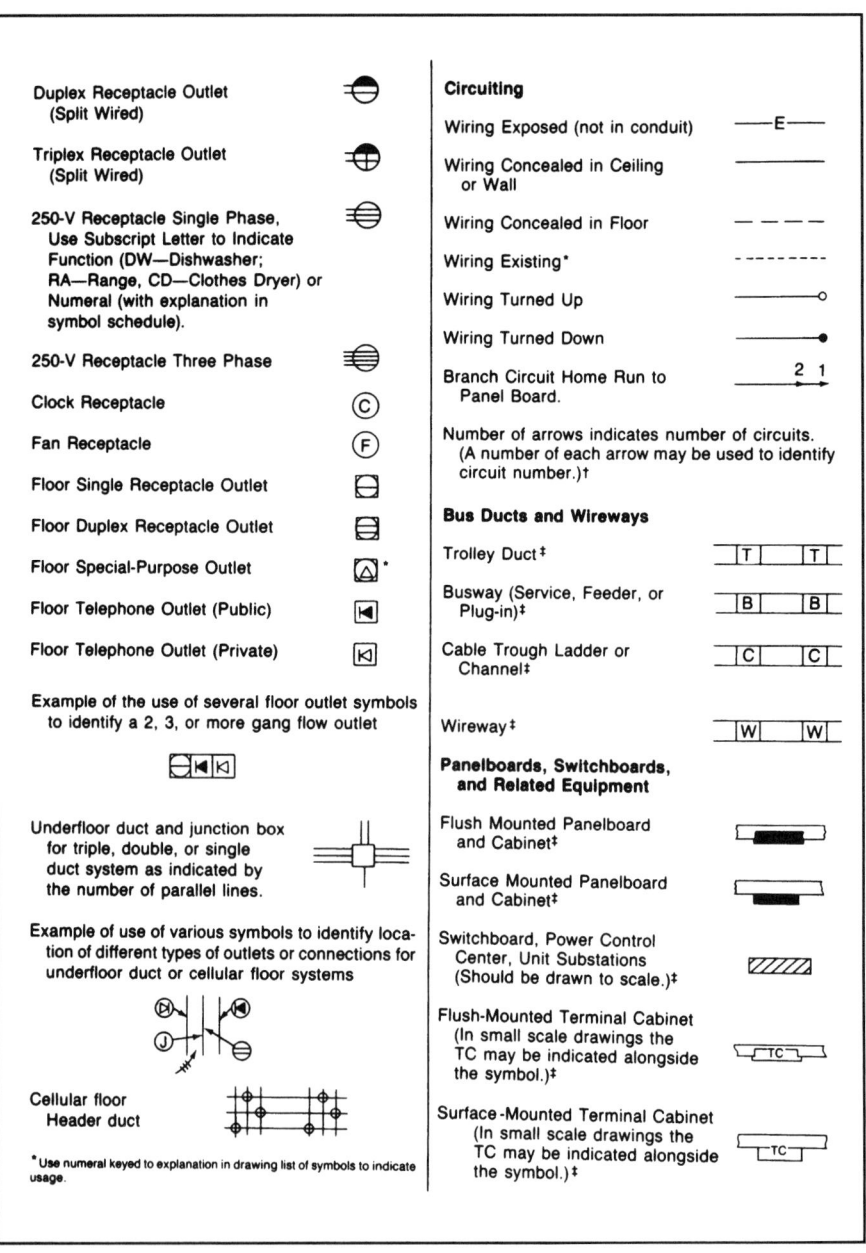

Fig. E-2: Electrical symbols (cont'd).

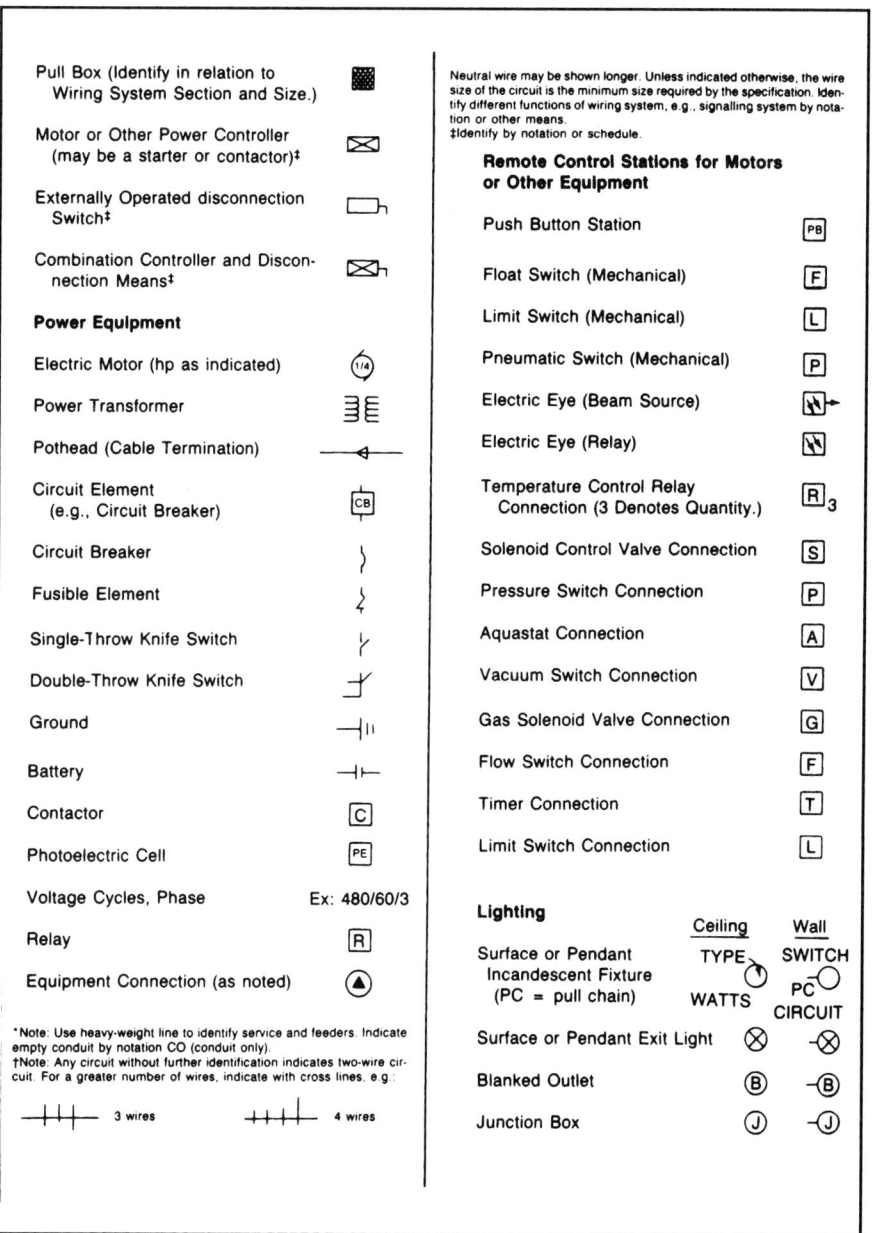

Fig. E-2: Electrical symbols (cont'd).

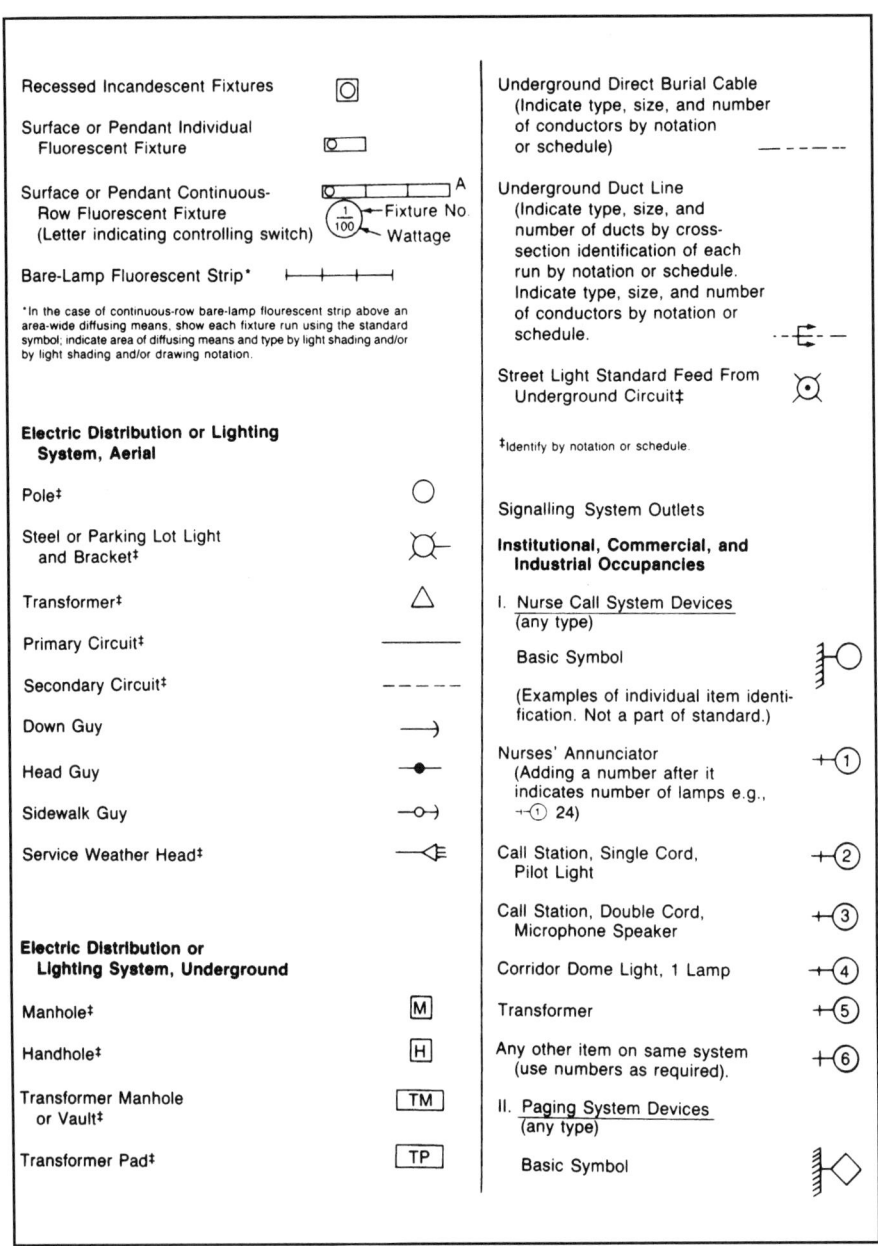

Fig. E-2: Electrical symbols (cont'd).

(Examples of individual item identification. Not a part of standard)

Keyboard

Flush Annunciator

Two-Face Annunciator

Any other item on same system - use numbers as required

III. Fire Alarm System Devices (any type) including Smoke and Sprinkler Alarm Devices

   Basic Symbol

   (Examples of individual item identification. Not a part of standard)

Control Panel

Station

10" Gong

Presignal Chime

Any other item on same system (use numbers as required).

IV. Staff Register System Devices (any type)

   Basic Symbol

   (Examples of individual item identification. Not a part of standard)

Phone Operators' Register

Entrance Register (flush)

Staff Room Register

Transformer

Any other item on same system (use number as required).

V. Electric Clock System Devices (any type)

   Basic Symbol

   (Examples of individual item identification. Not a part of standard)

Master Clock

12" Secondary (flush)

12" Double Dial (Wall Mounted)

18" Skeleton Dial

Any other item on same system (use numbers as required).

VI. Public Telephone System Devices

   Basic Symbol

   (Examples of individual item identification. Not a part of standard)

Switchboard

Desk Phone

Any other item on same system (use numbers required).

VII. Private Telephone System Devices (any type)

   Basic Symbol

   (Examples of individual item identification. Not a part of standard)

Switchboard

Wall Phone

Any other item on same system (use numbers as required).

VIII. System Devices (any type)

   Basic Symbol

   (Examples of individual item identification. Not a part of standard)

Fig. E-2: Electrical symbols (cont'd).

Fig. E-3: Several types of electric heating units used in residential applications.

electrolytic condenser-capacitor: Plate or surface capable of storing small electrical charges. Common electrolytic condensers are formed by rolling thin sheets of foil between insulating materials. Condenser capacity is expressed in microfarads.

electromagnet: A device consisting of a ferromagnetic core and a coil that produces appreciable magnetic effects only when an electric current exists in the coil.

electromagnetic force: A magnetic field that is created by an electrical force.

Effects of Electrical Shock		
	Current in milliamperes @ 60 Hz.	
	MEN	WOMEN
Cannot be felt	0.4	0.3
Felt slightly — mere tingle	1.1	.7
Shock — not painful	1.8	1.2
Shock— painful	9.0	6.0
Shock — painful; cannot let go, breathing stops	23.0	15.0
Ventricular fibrillation (usually fatal)	100	100

Fig. E-4: Electrocution. The stages of electrical shock.

electromagnetic interference (EMI): Electronic signals give off energy that is picked up by other electrical circuits.

electromechanical relay (EMRs): A device that converts electrical energy into mechanical switching.

electromotive force (emf) voltage: Electrical force that causes current (free electrons) to flow or move in an electrical circuit. Unit of measurement is the volt.

electron: The subatomic particle that carries the unit negative charge of electricity.

electronegative gas: A type of insulating gas used in pressure cables, such as SF6.

electron emission: The release of electrons from the surface of a material into surrounding space due to heat, light, high voltage, or other causes.

electronic ballast: The most energy efficient of all the ballasts. They do not contain coils, capacitors, or transformers and generally operate in the range of 25–35 kHz.

electronic control: A solid state device that controls components.

electronics: The science dealing with the development and application of devices and systems involving the flow of electrons in vacuum, gaseous media, and semi-conductors.

electron shells: The same as electron orbits. See Fig. A-12.

electro-osmosis: The movement of fluids through diaphragms because of electric current.

electroplating: Depositing a metal in an adherent form upon an object using electrolysis.

electropneumatic: An electrically controlled pneumatic device.

electroscope: An instrument designed to check the polarity of an electrostatic charge of an object.

electrostatic charge: A negative static charge exists when an object has a surplus of electrons. A positive static charge exists when an object has a deficiency of electrons.

electrostatic force: The force surrounding a charged object.

electrostatics: Electrical charges at rest in the frame of reference.

electrotherapy: The use of electricity in treatment of disease.

electrothermics: Direct transformations of electric and heat energy.

electrotinning: Depositing tin on an object.

element: There are over 100 different types of atoms and each type is called an element. Each element has its own set number of protons in each nucleus and each element has its own unique number of electrons surrounding its nucleus, which makes it different from all other elements.

elevation: Drawing showing the projection of a building on a vertical plane.

elevation drawings: Drawings that show the vertical dimensions of a building or dwelling.

elongation: 1) The fractional increase in length of a material stressed in tension. 2) The amount of stretch of a material in a given length before breaking.

EMA (Electrical Moisture Absorption): A water tank test during which the sample cables are subjected to voltage while the water is maintained at rated temperature; the immersion time is long, in order to accelerate failure due to moisture in the insulation; simulates buried cable.

EMF: Electromotive Force. The force or pressure that pushes electrons through a conductor.

EMI: Electromagnetic Interference.

emitter: The part of a transistor that emits electrons. See Fig. E-5.

EMT: Electrical Metallic Tubing, a name for thin-wall electrical tubing.

emulsifying agent: A material that increases the stability of an emulsion.

emulsion: The colloidal suspension of one liquid in another liquid, such as oil in water for lubrication.

enameled wire: Wire insulated with a thin baked-on varnish enamel, used in coils to allow the maximum number of turns in a given space.

enclosed: Surrounded by a case, housing, fence, or wall that will prevent persons from accidentally contacting energized parts.

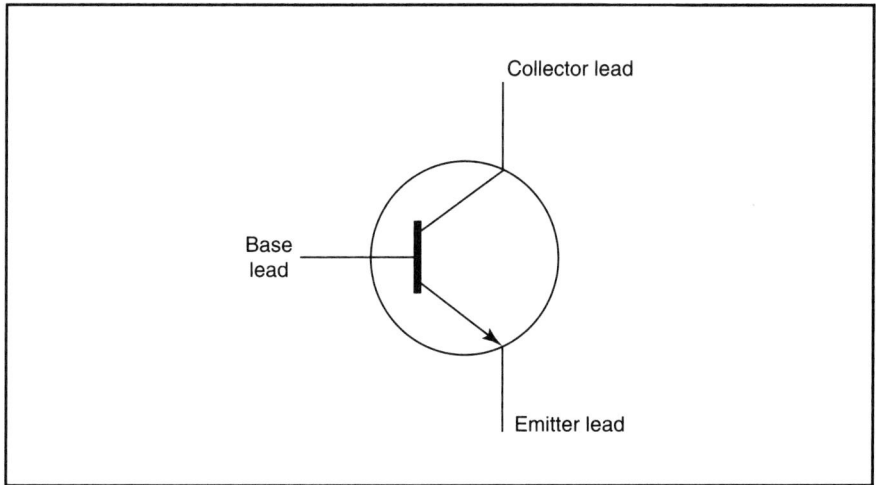

Fig. E-5: Emitter NPN symbol.

enclosure: The case or housing or apparatus, or the fence or walls surrounding an installation to prevent personnel from accidentally contacting energized parts, or to protect the equipment from physical damage.

end-of-the-run-wiring: An electrical or junction box that is at the end of the electrical circuit. Only one cable enters the end-of-the-run box.

end-wired switch: A switch that has its terminals at the top and/or bottom of the switch casing.

energizing current: The primary current in an unloaded transformer.

energy: The ability to do work; such as heat, light, electrical, mechanical, etc.

energy efficiency rating (EER): The amount of energy an appliance consumes. The higher the EER rating the more efficient the appliance.

energy-saving lamp: A lamp that produces more lumens with less watts.

energy savings: The use of energy-efficient appliances and lights will bring energy savings to the consumer.

engine: An apparatus that converts heat to mechanical energy.

ENT: Electrical Nonmetallic Tubing.

entrance cap: The point where the service loop ends and the building service begins. This cap is weather proof and generally furnished by the electrical contractor. See Fig. E-6.

entrance ell: A 90° fitting that connects to the service entrance conduit and gives access so that the service conductors may be pulled from either direction.

environment: 1) The universe within which a system must operate. 2) All the elements over which the designer has no control and that affect a system or its inputs.

EPA (Environmental Protection Agency): The federal regulatory agency responsible for maintaining and improving the quality of our living environment—mainly air and water.

Equilibrium

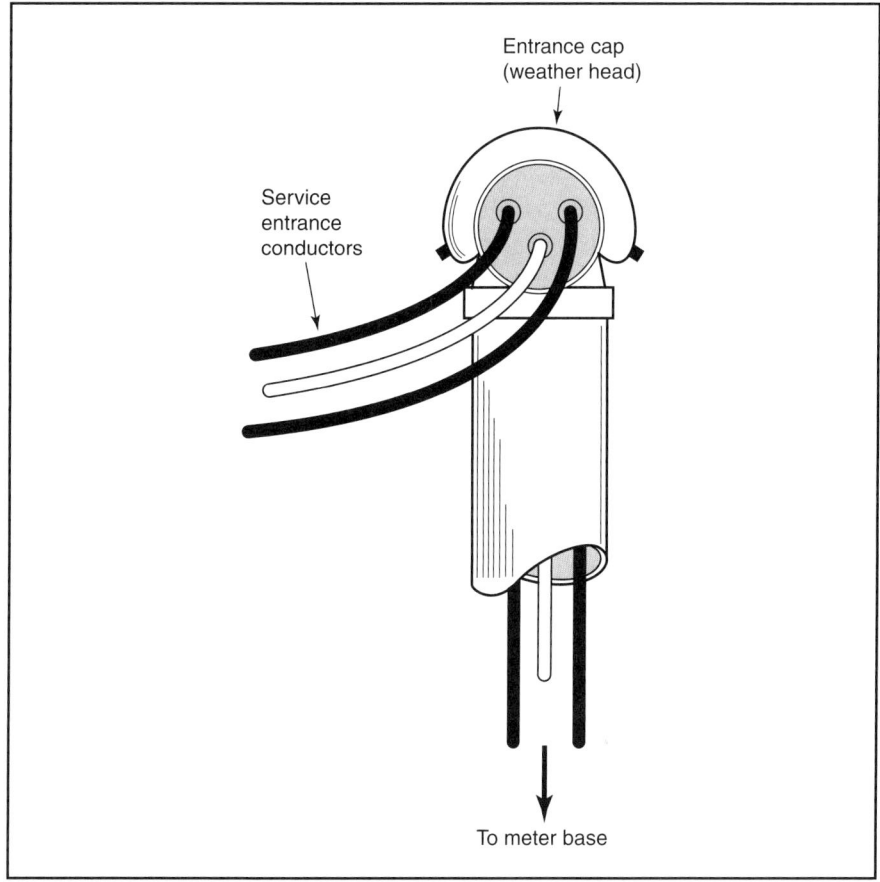

Fig. E-6: Entrance cap. A weatherproof cap used on an overhead entrance service.

epitaxial: A very significant thin-film type of deposit for making certain devices in microcircuits involving a realignment of molecules.

EPRI (Electric Power Research Institute): An organization to develop and manage a technology for improving electric power production, distribution, and utilization; sponsored by electric utilities.

equilibrium: Properties are time constant.

equipment: A general term including material, fittings, devices, appliances, fixtures, apparatus, and the like used as part of, or in connection with, an electrical installation.

equipment grounding conductor: The conductor used to connect the noncurrent-carrying metal parts of equipment, raceways, and other enclosures to the system grounded conductor and/or the grounding electrode conductor at the service equipment or at the source of a separately derived system.

equipotential: Having the same voltage at all points.

equivalent circuit: An arrangement of circuit elements that has characteristics over a range of interest electrically equivalent to those of a different circuit or device.

ERDA (Energy Research & Development Administration): Federal agency (replacing part of AEC) for research and relating to energy—new sources, better efficiency, etc.

erosion: Destruction by abrasive action of fluids.

escutcheon: The plate about a keyhole or the one to which a door knocker is attached.

estimating: Calculating the amount of material required for a piece of work; also the labor required and determining the value of the finished product.

etching: Revealing structural details of a metal surface using chemical or electrolytic action.

ETL: Electrical Testing Laboratory.

eutectic alloy: A metal with a very fast and low melting point. Used in devices that use thermal overloads.

evaporation: A term applied to the changing of a liquid to a gas; heat is absorbed in this process.

evaporator: Part of a refrigerating mechanism in which the refrigerant vaporizes and absorbs heat.

excavation: A digging out of earth to make room for engineering improvements. A cavity is so formed.

excitation losses: Losses in a transformer or electrical machine because of voltage.

excite: To initiate or develop a magnetic field.

exclusive OR gate: A logic gate that produces an output only when its inputs have opposite logic.

expansion bolt: Bolt with a casing arranged to wedge the bolt into a masonry wall to provide an anchor.

expansion joint: Joint between two adjoining concrete members arranged to permit expansion and contraction with temperature changes.

expansion, thermal: The fractional change in unit length per unit temperature change.

expansion valve: A device in a refrigerating system that maintains a pressure difference between the high side and low side and is operated by pressure.

explosionproof: Designed and constructed to withstand an internal explosion without creating an external explosion or fire. Several explosionproof fittings are shown in Fig. E-7.

Fig. E-7: Several types of explosionproof fittings.

explosionproof apparatus: Apparatus enclosed in a case that is capable of withstanding an explosion of a specified gas or vapor that may occur within it, and also capable of preventing the ignition of a specified gas or vapor surrounding the enclosure by sparks, flashes, or explosion of the gas or vapor within, and which operates at such an external temperature that a surrounding flammable atmosphere will not be ignited thereby.

exponential: Pertaining to the mathematical expression, $y = ae^{bx}$.

exposed (as applied to live parts): Live parts that a person could inadvertently touch or approach nearer than a safe distance. This term is applied to parts not suitably guarded, isolated, or insulated.

exposed (as applied to wiring method): Not concealed; externally operable; capable of being operated without exposing the operator to contact with live parts.

extender: A substance added to a plastic to reduce the amount of the primary resin required per unit area.

exterior: The outside of, as a whole or in part, as the exterior of a building; an exterior wall.

externally operable: Capable of being operated without exposing the operator to contact with live parts.

extraction: The transfer of a material from a substance to a liquid in contact with the substance.

extrude: To form materials to a given cross section by forcing through a die.

extruder types: 1) Strip—uses strips of compound. 2) Powder/pellet—uses compound in powder or pellet form.

eyelet: Something used on printed circuit boards to make reliable connections from one side of the board to the other.

F

fabrication: The act of building or putting together. Forming into a whole by uniting or assembling of parts.

facade: Main front of a building.

face: An operation that machines the sides or ends of the piece.

face brick: Brick selected for appearance in an exposed wall.

face of a gear: That portion of the tooth curve above the pitch circle and measured across the rim of the gear from one end of the tooth to the other.

facsimile: The remote reproduction of graphic material; an exact copy.

factorial experiment: Having more than one factor as a controlled variable in one experiment; produces much data per experiment, but the results are complex to analyze.

factor of safety: Ratio of ultimate strength of material to maximum permissible stress in use.

Fahrenheit: The commonly used thermometer scale named after Gabriel Fahrenheit, a German physicist (1686–1736); the freezing point is 32 degrees, and the boiling point 212 degrees.

fail-safe control: A device that opens a circuit when the sensing element fails to operate.

failure: Termination of the ability of an item to perform the required function.

fan: A radial or axial flow device used for moving or producing artificial currents of air.

fan control: A variable-speed device that controls fans.

FAO: This symbol on a mechanical drawing means that the piece is machined or finished all over.

farad: The basic unit of capacitance; 1 farad equals one coulomb per volt: $1f = m^{-2}.kg^{-1}s^4.A^2$. It is too large a unit for practical work. Its common use is in terms of "microfarad."

Faraday, Michael: (1791–1867). English scientist; early investigator and experimenter who invented the transformer.

Faraday's laws of electrolysis: 1) The weight of the products of electrolysis is proportional to the quantity of electricity that has passed through the electrolyte. 2) For a given quantity of electricity the weight of the products of electrolysis is proportional to their electrochemical equivalents.

fatigue: The weakening or breakdown of a material due to cyclic stress.

fatigue strength: The maximum stress that can be sustained for a specified number of cycles without failure, the stress being completely reversed within each cycle unless otherwise stated.

fault: An abnormal connection in a circuit.

fault, arcing: A fault having high impedance causing arcing.

fault, bolting: A fault of very low impedance.

fault current: Any current that flows outside of its intended path.

fault, ground: A fault to ground.

FCC: Flat Conductor Cable. Type FCC consists of three or more flat copper conductors placed edge-to-edge and separated and enclosed within an insulating assembly.

Federal Communications Commission (FCC): A federal government agency established to regulate interstate communications.

feedback: The process of transferring energy from the output circuit of a device back to its input.

feeder: A circuit, such as conductors in conduit or a busway run, which carries a large block of power from the service equipment to a sub-feeder panel or a branch circuit panel or to some point at which the block or power is broken down into smaller circuits.

feedhorn: The device that is mounted on the front of the satellite dish that downlinks the signal. See Fig. F-1.

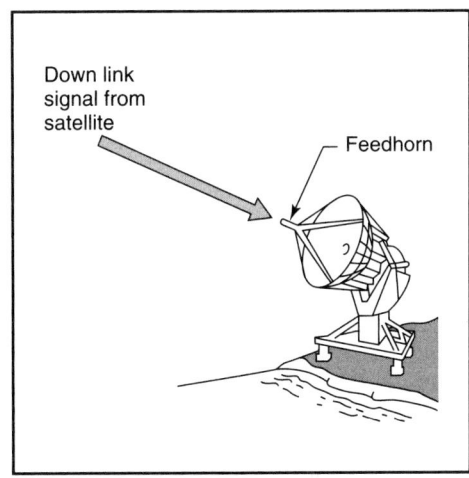

Fig. F-1: Feedhorn (satellite dish) used to downlink a signal from a satellite.

feed wire: The hot wire that brings power into electrical boxes.

feeler gauge: A tool that has thin, flat blades that measure clearances in thousandths of an inch.

Ferranti effect: When the voltage is greater than the source voltage in an ac cable or transmission line.

festoon lighting: A string of outdoor lighting that stretches more than fifteen feet.

fiber optics: Transmission of energy by light through glass fibers.

fibers (rope): Materials from which ropes are made.

fibrillation: A continued, uncoordinated activity in the fibers of the heart, diaphragm, or other muscles.

fiddle: A small, hand-operated drill.

fidelity: The degree to which a system accurately reproduces an input.

field: The effect produced in surrounding space by an electrically charged object, by electrons in motion, or by a magnet.

field coil: The coil or winding around the field magnets or pole pieces of a motor or generator.

field core: The iron projection usually salient, upon which is wound the field winding of a generator or motor.

field density: The density of the magnetic field or the magnetic flux, measured in the number of lines of force per unit area, is dependent upon the strength of the field element, the number of turns of wire, and the size and characteristics of the pole piece.

field distortion: The distortion of the normal field existing between the north and south poles of a generator due to the counter electromotive force generated in the armature windings.

field, electrostatic: The region near a charged object.

field excitation: The magnetic effect produced in an electromagnet when current is passed through a winding, usually with an iron core.

field magnet: The electromagnet by which the magnetic field of force is produced in a generator.

field rheostat: A variable high resistance of low-current capacity inserted in the field circuit to regulate within limits the output of a generator.

field winding: Two different types of field windings—shunt and series—are used in dc motors.

filament: A cathode in the form of a metal wire in an electron tube; a thin wire or fiber.

filled tape: A fabric tape having interstices, but not necessarily the surface, filled with a compound to prevent wicking, improve strength, make conductive, etc.

filler: A cheap and relatively inert substance added to plastic or rubber to make it less costly and improve physical properties.

filler, cable: Materials used to fill voids and spaces in a cable construction; normally to give a smooth outer configuration, and also may serve as flame retardants, etc.

fillet: The rounded corner or portion that joins two surfaces at an angle to each other.

film: A rectangular product having thickness of 0.010 inch thick or less.

filter: A porous article through which a gas or liquid is passed to separate out matter in suspension; a circuit or devices that pass one frequency or frequency band while blocking others, or vice versa.

final: The final inspection of an electrical installation, e.g., "The contractor got the final on the job."

final tests: Those performed on the completed cable (after manufacturing).

fine print note: NEC®, informational only and not enforceable.

fines: Fill material such as rocks having ⅛ inch as the largest dimension.

finish plaster: Final or white coat of plaster.

firebrick: Brick made to withstand high temperatures and used for lining chimneys, incinerators, and similar structures.

fireproof wood: Chemically treated wood; fire-resistive, used where incombustible materials are required.

fire-rated doors: Doors designed to resist standard fire tests and labeled for identification.

fire-resistance rating: The time in hours the material or construction will withstand fire exposure as determined by certain standards.

fire-shield cable: Material or devices to prevent fire spread between raceways.

fire-stop: A barrier to minimize fire spread.

fish: To fish wire or cable means to pull it through conduit, raceway or other confined spaces, like walls or ceilings.

fish tape: A flexible metal tape for fishing through conduits or other raceway to pull in wires or cables; also made in non-metallic form of "rigid rope" for hand fishing of raceways. See Fig. F-2, on page 98.

fishing chain (drop chain): A method of using a light furnace type chain to fish down between walls to pull wire back through after connecting to the chain.

fission: (nuclear power): The splitting of an atom into two fragments by bombarding its nucleus with particles releasing high kinetic energy (32pj) and two or three neutrons along with radiation; the most important type of fission is that caused by neutrons because it can be self-sustaining due to chain reactions—the newly released neutrons can cause other fissions to occur.

fitting: An accessory such as a locknut, bushing, or other part of a wiring system that is intended primarily to perform a mechanical rather than an electrical function. See Fig. F-3.

555 timer: A versatile eight-pin integrated circuit timer. Used where time delay is an important factor and also as an oscillator where electronic pulses are used to operate computer circuits.

five hundred thousands: Referring to size of conductors by their MCM or current kcmil rating, e.g., "Two hundred and fifty thousands" is a

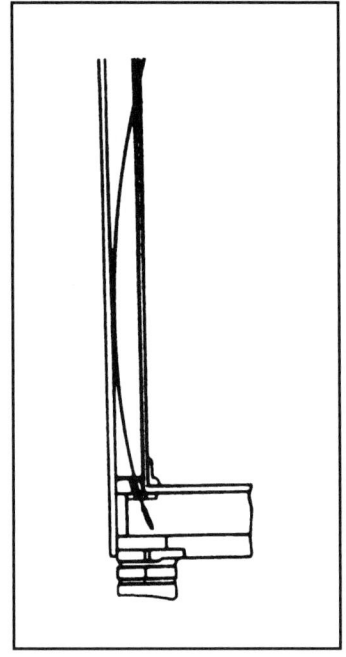

Fig. F-2: Fish tape used to pull wires through a partition.

Fig. F-3: Several types of fittings that are used in a mechanical way rather than electrically on electrical systems.

number of 250 kcmil conductors; "Twin three hundred thousands" is two conductors of 300,000 circular mil size. The *NEC*® now refers to kcmil instead of the former MCM.

fixed rate-of-rise detectors: A device that affects the operating element when a rapid rise in temperature is detected.

fixed resistor: A resistor that has a fixed resistance and cannot be altered.

fixture: The complete assembly of components that enables permanent attachment of the fixture.

fixture schedule: A schedule that is a part of the electrical blueprints and/or specifications explaining the type of fixture, identification mark, and where it is to be installed.

fixture splice: The fixture wire is wound in close turns around the main conductor after which the end of the main conductor is bent tightly back over the coils of fixture wire.

fixture stud: A special fitting that connects the light fixture to an electrical box. See Fig. F-4, page 100.

fixture whips: Flexible prewired cables attached to light fixtures for easy connection to the electrical circuit.

fixture wire: Usually 16 or 18 gauge, solid or stranded and insulated. It is used for wiring electric fixtures.

flag: A visual indicator for event happenings such as the activation and reclosing of an automatic circuit breaker.

flame detector: A device that detects the radiant energy emitted by a flame.

flame-retardant: 1) Does not support or convey flame. 2) An additive for rubber or plastic that enhances its flame resistance.

flange: The circular disks on a reel to support the drum and keep the cable on the drum.

flanged box: Junction box connected to an open ceiling joist.

flapper valve: The type of valve used in refrigeration compressors that allow gaseous refrigerants to flow in only one direction.

flasher: A device for periodically flashing on and off a lamp or group of lamps, as in some electric signs.

Fig. F-4: A fixture stud connects a fixture to an electrical box.

flashover: A momentary electrical interconnection around or over the surface of an insulator.

flashpoint: The lowest temperature at which a combustible substance ignites in air when exposed to flame.

flat: Of uniform thickness; eliminates the drops of beams and girders.

flat wire: A rectangular wire having 0.188 inch thickness or less, 1¼-inch width or less.

Fleming's rule: Right-hand rule. If the forefinger points along the lines of flux, and the thumb in the direction of the motion of the conductor, the middle finger will point in the direction of the induced e.m.f. Left-hand rule: Point the forefinger in the direction of the flux, the middle finger in the direction of the current in the conductor, then the thumb will point in the direction in which the conductor tends to move. Fleming's rules determine the direction in which a motor will rotate, or the polarity of a generated current from a generator.

Flemish bond: Pattern of bonding in brickwork consisting of alternate headers and stretchers in the same course.

flex: Common term used to refer to flexible metallic conduit. See *Greenfield*.

flexural strength: The strength of a material in bending, expressed as the tensile stress of the outermost fibers of a bent test sample at the instant of failure.

flitch beam: Built-up beam consisting of a steel plate sandwiched between wood members and bolted.

floating: Not having a distinct reference level with respect to voltage measurements. For example, a "floating neutral" on a balanced three-wire delta system.

float switch: A switch that is opened and closed by a float that rises and falls with the level of the liquid in a tank. See Fig. F-5.

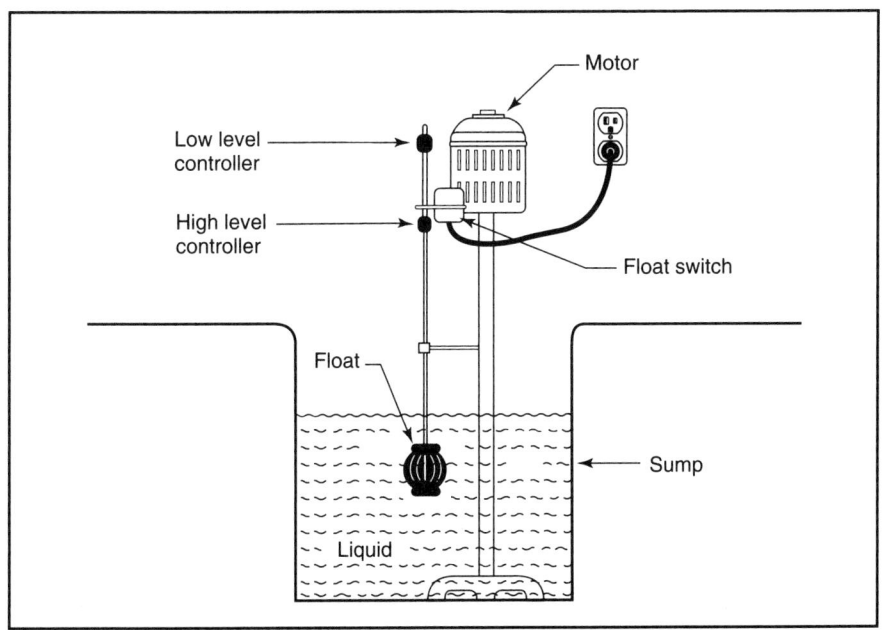

Fig. F-5: A float switch controls the level of tanks or sumps.

float valve: Type of valve that is opened by a sphere or pan that floats on a liquid surface and controls the level of liquid.

flooding: Act of filling a space with a liquid.

floor plan: A top view plan of the layout of a building's rooms, doors, windows, and their sizes.

flow meter: Instrument used to measure velocity or volume of fluid movement.

flow switch: A switch that detects the flow of fluidics.

fluctuating dc: A dc that has a varying voltage (amplitude) but never reaches zero voltage.

fluctuation: A variation, an irregular change of movement.

fluorescent lamp: A lamp that has a filament at each end that gives off an electric charge through low-pressure mercury and causes the coating on the glass tubing to produce a light.

flux: 1) The rate of flow of energy across or through a surface. 2) A substance used to promote or facilitate soldering or welding by removing surface oxides.

flux density: Amount of flux per cross sectional area of magnetized material.

flux leakage: Flux lines that do not follow the iron core of a generator and are of no value.

foamed insulation: Insulation made sponge-like by using foaming or blowing agents to create the cells.

foil: Metal film.

folding wood rule: A folding rule that, when fully extended, is six feet long. It measures in $1/16$-inch increments and generally has a stud marking for every 16 inches. See Fig. F-6 on page 103.

footcandle: One footcandle equals one lumen per square foot.

footing: Structural unit used to distribute loads to the bearing materials.

footlambert: Reflected light from a diffusing surface at the rate of one lumen per square centimeter.

foot switch: A control operated by a person's foot.

Friction 103

Fig. F-6: A folding rule extends to six feet, and generally it has a stud marking every 16 inches.

force: Motor torque needed to turn a motor.

forced convection: Movement of fluid by mechanical force such as fans or pumps.

forward-biased voltage: The voltage or current that turns on the electronic device.

foundation: Composed of footings, piers, foundation walls (basement walls), and any special underground construction necessary to properly support the structure.

foundation plan: The sectional view that shows the footing, foundation walls, and all structural details needed to build a building's foundation.

four-way switch: A switch that is used in conjunction with two three-way switches when a load has to be turned on or off at more than two locations. This switch is connected to the travelers of the three-way switches. See Fig. F-7 on page 104.

FPM: Feet Per Minute.

FR-1: See *VW-1*.

free electrons: Valence electrons that are not connected to any atoms.

frequency: The number of complete cycles an alternating electric current, sound wave, or vibrating object undergoes per second.

friction: Any resistance to motion.

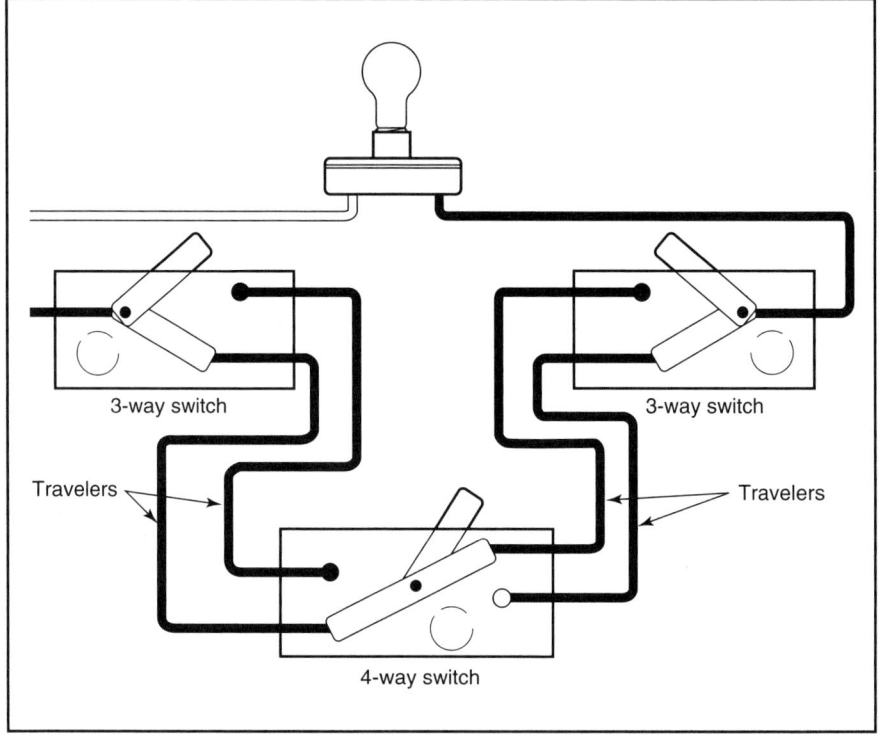

Fig. F-7: A four-way switch must be used with two three-way switches.

friction pile: Pile with supporting capacity produced by friction with the soil in contact with the pile.

friction tape: An insulating tape made of asphalt impregnated cloth; used on 600V cables.

front-wired switch: The terminals are mounted on the front or face of the switch.

frost line: Deepest level below grade to which frost penetrates in a geographic area.

fuel cell: A cell that can continually change chemical energy to electrical energy.

full braid: One made of a single material as opposed to one of a mixture of materials.

full load current rating: The current the motor is expected to need when running at certain speeds, voltages, and/or loads.

full load torque: The torque required to produce the rated power of a motor at a rated speed.

$$T = HP \times 5252 \div rpm$$

full voltage control: Equipment that is connected to the line that supplies power when starting.

fully rated system: See *series rated system.*

function: A quantity whose value depends upon the value of another quantity.

functional light: Lighting that illuminates stairways, walkways, gates, and any area where safety is a high priority.

furring: Thin wood, brick, or metal applied to joists, studs, or wall to form a level surface (as for attaching wallboard) or airspace.

fuse: A protecting device that opens a circuit when the fusible element is severed by heating, due to overcurrent passing through. Rating: voltage, normal current, maximum let-thru current, delay of interruption.

fuse box: A metal cabinet that holds fuses and a neutral bus bar.

fuse clips: The spring part of a cutout or switch that holds the ferrules of a cartridge fuse.

fuse, dual element: A fuse having two fuse characteristics; the usual combination is having an overcurrent limit and a time delay before activation. See Fig. F-8 on page 106.

fuse (enclosed cartridge): Designed to protect a single major piece of equipment or appliance. It is cylindrical in shape and can handle higher voltages than the normal plug fuses.

fuse link: The fusible part of a cartridge fuse.

fuse, nonrenewable or one-time: A fuse that must be replaced after it interrupts a circuit.

fuse plug: A fuse that generally comes in sizes Type S and time-de-lay. A metal wire melts and opens the circuit when a short occurs in the electrical system.

Fig. F-8: Fuses are manufactured in a variety of shapes and sizes: 1) Edison-base plug fuse; 2) type S plug fuse; 3) type S plug fuse adapter; 4) knife-type cartridge fuse with renewable element; 5) ferrule-type cartridge fuse with non-renewable element.

fuse, renewable link: A fuse that may be reused after current interruption by replacing the meltable link.

fuse (S type): Plug type fuses that are not rated over 125 volts and 30 amps that have non-interchangeable bases.

fuse (time delay): Allows for a momentary power surge when equipment first turns on or is cycling without causing the fuse to open.

fuse wire: Wire made of an alloy that melts at a low temperature.

fusible plug: A plug or fitting made with a metal of a known low melting temperature; used as a safety device to release pressures in case of fire.

G

gable: The triangular end of an exterior wall above the eaves.

gable roof: A ridge roof terminating in a gable.

gain: 1) The ratio of output to input power, voltage, or current, respectively. 2) The increase in signal power produced by an amplifier. 3) When bending pipe, the distant of the long side of the bend compared to the short side of the bend is called gain.

galvanic action: The action upon one another of electropositive and electronegative metals, wasting away the positive metal; particularly noticeable where iron and copper, or zinc and copper, or brass and iron, are in contact in the presence of acidulated water.

galvanometer: An instrument for indicating or measuring a small electrical current by means of a mechanical motion derived from electromagnetic or dynamic forces.

gambrel roof: Roof with its slope broken by an obtuse angle.

gangable boxes: Electrical boxes with removable sides that are designed to be connected together in a gang.

gang switch: A unit of two or more switches to give control of two or more circuits from one point. The entire mechanism is mounted in one box under one cover.

garage: A building or portion of a building in which one or more self-propelled vehicles carrying volatile, flammable liquid for fuel or power are kept.

garden bond: Bond formed by inserting headers at wide intervals.

gas: Vapor phase or state of a substance.

gas filled pipe cable: See *pressure cable.*

gas pocket: A cavity caused by entrapped gas.

gate: A device with multiple inputs and a single output that makes an electronic circuit operable for a short time. There are five basic types of gates: AND, OR, NAND, NOR, and inverter.

gauge: 1) Dimension expressed in terms of a system of arbitrary reference numbers; dimensions expressed in decimals are preferred. 2) To measure.

gem box: The most common rectangular outlet box used to hold wall switches and receptacle outlets installed recessed in walls; made in wide variety of constructions: 2 in. wide by 3 in. high by various depths; without clamps for conduit and with or without clamps for cable (armored or nonmetallic sheathed); single-gang boxes that can be ganged together for more than one device.

general alarm: An alarm that warns the occupants of a building of impending danger without identifying the exact cause or location.

general lighting: Lighting designed to illuminate an entire room. Different from task lighting or outline lighting.

general-purpose motor: A motor that has a continuous rating of 40°C and is designed to have standard operating characteristics without being restricted to a particular type of application. (NEMA)

generator: 1) A rotating machine that is used to convert mechanical to electrical energy. 2) Automotive-mechanical to direct current. 3) General-apparatus, equipment, etc. to convert or change energy from one form to another.

geometric factor: A parameter used and determined solely by the relative dimensions and configuration of the conductors and insulation of a cable.

GFCI (Ground-Fault Circuit-Interrupter): A protective device that detects abnormal current flowing to ground and then interrupts the circuit.

GFPE (Ground-Fault Protection of Equipment): A system intended to provide protection of equipment from damaging line-to-ground fault currents by operating to cause a disconnecting means to open all ungrounded conductors of the faulted circuit.

giga: Equals one billion.

gigahertz (Ghz): One billion cycles per second.

girder: A large beam made of wood, steel, or reinforced concrete.

girt: Heavy timber framed into corner posts as support for the building.

government anchor: A V-shaped anchor usually made of ½-inch round bars to secure the steel beam to masonry.

GR: Terminal screw for a ground wire.

grade beam: Horizontal, reinforced concrete beam between two supporting piers at or below ground supporting a wall or structure.

graded insulation: Combining insulations in a manner to improve the electric field distribution across the combination.

gradient: The rate of change of a variable magnitude.

grain: An individual crystal in a polycrystalline metal or alloy.

gray: The derived SI unit for absorbed radiation dose; one gray equals one joule per kilogram.

Greenfield: Another name used to refer to flexible metal conduit.

green wire: The color of a conductor coating that indicates that a conductor is to be used for grounding only. (*NEC®*)

grid: An electrode having one or more openings for the passage of electrons or ions.

grid leak: A resistor of high ohmic value connected between the control grid and the cathode in a grid-leak capacitor detector circuit and used for automatic biasing.

grillage: Steel framework in a foundation designed to spread a concentrated load over a wider area; generally enclosed in concrete.

grille: An ornamental or louvered opening placed at the end of an air passageway.

groined ceiling: Arched ceiling consisting of two intersecting curves, planes.

grommet: A plastic, metal, or rubber doughnut-shaped protector for wires or tubing as they pass through a hole in an object.

ground: A large conducting body (as the earth) used as a common return for an electric circuit and as an arbitrary zero of potential.

ground check: A pilot wire in portable cables to monitor the grounding circuit.

ground clamp: A clamp used for attaching a wire or other conductor to a pipe to make a good electrical connection.

ground clip: A spring clip used to secure a bonding conductor to an outlet box. See Fig. G-1.

ground coil: A heat exchanger buried in the ground that may be used either as an evaporator or a condenser.

ground conductor: A conductor used to connect equipment or the grounded circuit of a wiring system to a grounding electrode or electrodes.

grounded: Connected to earth or to some conducting body that serves in place of the earth.

Fig. G-1: A ground clip and its practical applications.

grounded conductor: A system or circuit conductor that is intentionally grounded, usually gray or white in color.

ground fault: A leakage or loss of current to ground.

ground fault detector: A fire system that has a ground-fault monitoring circuit.

ground-fault protection of equipment: See *GFPE*.

grounding: The device or conductor connected to ground designed to conduct only in abnormal conditions.

grounding conductor: A conductor used to connect metal equipment enclosures and/or the system grounded conductor to a grounding electrode, such as the ground wire to the water pipe at a service; also may be a bare or insulated conductor used to ground motor frames, panel boxes, and other metal equipment enclosures used throughout an electrical system. In most conduit systems, the conduit is used as the ground conductor. See Fig. G-2.

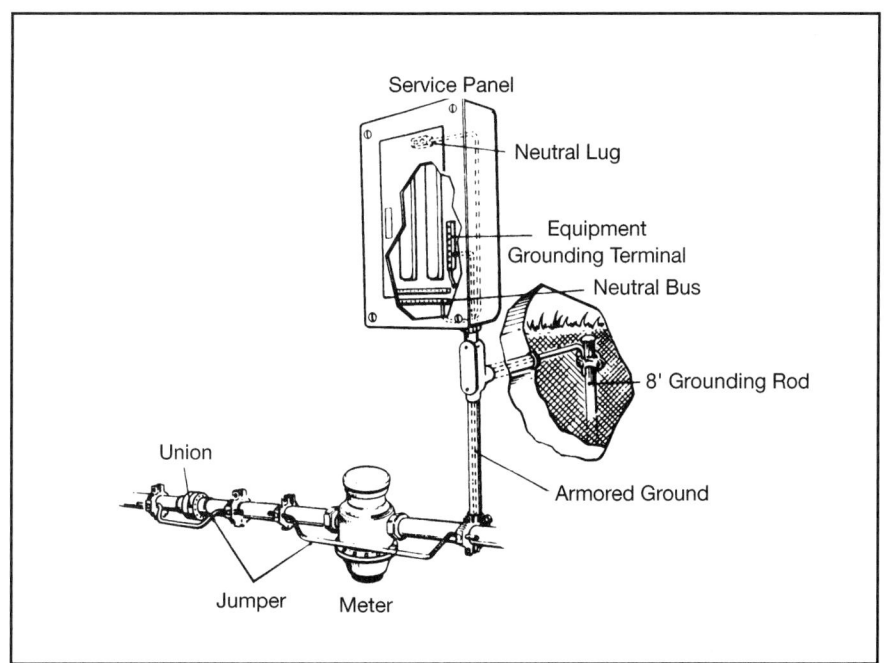

Fig. G-2: One *NEC*®-approved method of grounding an electric service.

grounding electrode (rod) conductor: The conductor that ties the grounding electrode (rod) to the grounded conductor at the service or the equipment grounding conductor. See Fig. G-2.

grounding rod: Generally a driven rod ⅝ inch in diameter and 8 feet long used for connecting the grounding conductor. See Fig. G-2.

grounds: Narrow strips of wood nailed to walls as guides to plastering and as a nailing base for interior trim.

ground-wired receptacle: A receptacle that has a green grounding screw in addition to the hot and neutral lugs. The face of the receptacle has three openings, one for the neutral plug, one for the hot plug, and one for the ground prong.

group ambient temperature: The no-load temperature of a cable group with all other cables or ducts loaded.

guard: 1) A conductor situated to conduct interference to its source and prevent its influence upon the desired signal. 2) Covered, shielded, fenced, enclosed, or otherwise protected by means of suitable covers, casings, barriers, rails, screens, mats, or platforms to remove the likelihood of approach or contact by persons or objects to a point of danger.

guider: The adjustable part of the mold of an extruder.

gusset: A plate or bracket for strengthening an angle in framework.

gutter: The space provided along the sides and at the top and bottom of enclosures for switches, panels, and other apparatus, to provide for arranging conductors that terminate at the lugs or terminals of the enclosed equipment. Gutter is also used to refer to a rectangular sheet metal enclosure with removable side, used for splicing and tapping wires at distribution centers and motor control layouts. Usually called "auxiliary gutter."

guy: A tension wire connected to a tall structure such as a power pole and another fixed object (anchor) to add strength to the structure.

habitable room: A space for living, eating, sleeping, or cooking, according to the *NEC®*.

hack saw: A light-framed saw used for cutting metal conduit, BX cable, and other metal objects.

HACR: A tested circuit breaker approved for heating and air conditioning and refrigeration equipment.

half hard: A relative measure of conductor temper.

half lap, joint: Joint formed by cutting away half the thickness of each piece.

half wave: Rectifying only half of a sinusoidala ac supply. See Fig. H-1.

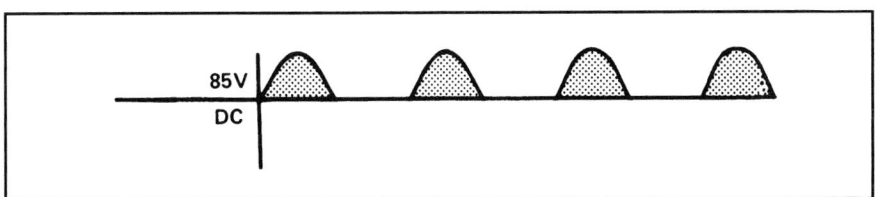

Fig. H-1: Half-wave depicted graphically.

half-wave rectifier: A device that changes ac to dc and blocks one half of the wave form, creating pulsating dc.

Hall effect: The changing of current density in a conductor due to a magnetic field extraneous to the conductor.

halogen: "Salt former." Applied to the family of elements consisting of fluorine, chlorine, bromine, and iodine.

halogen lamp: A high-pressure incandescent lamp that contains gases, which allows for a higher-efficiency lamp.

hammer drill: A drill that has a hammer and rotary action at the same time. Special bits must be used for this type of action.

handhole: A small box in a raceway used to facilitate cable installation into which workers reach but do not enter. See Fig. H-2.

handy box: The commonly used, single-gang outlet box used for surface mounting to enclose receptacles or wall switches on concrete or cinder block construction of industrial and commercial buildings; non-gangable; also made for recessed mounting; also known as "utility boxes." See Fig. H-3 on page 115.

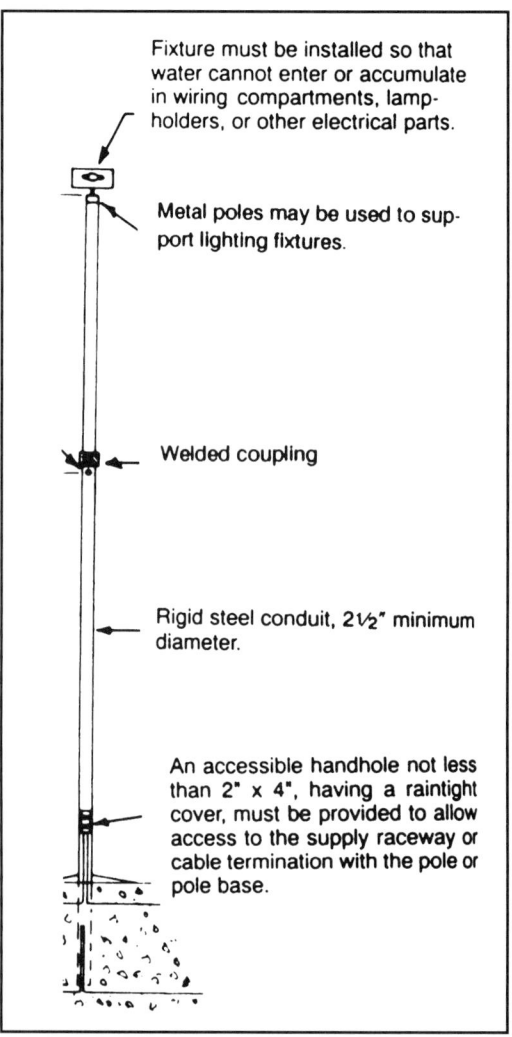

Fig. H-2: Handhole in Lighting Standard to facilitate wire connections.

hard drawn: A relative measure of temper; drawn to obtain maximum strength.

hardness: Resistance to plastic deformation usually by deformation; stiffness or temper; resistance to scratching, abrasion or cutting.

harmonic: An oscillation whose frequency is an integral multiple of the fundamental frequency.

harness: A group of conductors laced or bundled in a given configuration, usually with many breakouts.

harp: The metal arch that is used to support a lamp shade.

Fig. H-3: Utility or "handy" box

hat: A special pallet for transporting long rubber strips or coils of wire; the pallets look like a hat.

hazardous (classified location): A location in which ignitable vapors, dust, or fibers may cause fire or explosion.

HAZCOM: The term used for communication about hazards established by the Federal Government to warn workers of hazardous materials in use in their work area.

H beam: Steel beam with wider flanges than an I beam.

HDP: High Density Polyethylene.

header: 1) Brick laid with an end exposed in the wall. 2) Wood beam set between two trimmers and carrying the tail beams. 3) A transverse raceway for electric conductors, providing access to predetermined cells of a cellular metal floor, thereby permitting the installation of electric conductors from a distribution center to the cells.

heat: A fundamental physical force or energy relating to temperature.

heat dissipation: The flow of heat from a hot body to a cooler body by convection, radiation, or conduction.

heater: 1) A device that is designed to raise the temperature of the surrounding area. 2) An element within the overload protection on motor starters.

heat exchanger: A device used to transfer heat from a warm or hot surface to a cold or cooler surface. Evaporators and condensers are heat exchangers.

heat load: Amount of heat, measured in Btu, that is removed during a period of 24 hours.

heat pump: A compression cycle system used to supply heat to a temperature-controlled space, which can also remove heat from the same space.

heat sink: A part used to absorb heat from another device.

heat transfer: Movement of heat from one body or substance to another. Heat may be transferred by radiation, conduction, convection, or a combination of these.

heat treatment: Heating and cooling a solid metal or alloy to obtain desired properties or conditions; excluding heating for hot work.

heating cables: A type of resistance-type cable that can be installed in plaster walls to heat a room, and outside on eave spouts or in driveways to help melt the ice and snow.

heating element: A resistor that heats up when current flows through it and the heat is transferred to the surrounding area.

heating unit: The part of any electrical heating device in which the heat is generated.

heating value: Amount of heat that may be obtained by burning a fuel; usually expressed in Btu per pound or gallon.

heavy water (nuclear power): Heavy water, D^2O, contains deuterium which is hydrogen atoms having twice the ordinary mass.

helix: The path followed when winding a wire or strip around a tube at a constant angle.

henry: The derived SI unit for inductance; one henry equals one weber per ampere. A circuit has an inductance of one henry when a current changing at the rate of one ampere per second induces one volt in the circuit.

hermetic motor: A motor designed to operate within refrigeration fluid. See *NEC® 440.2*.

hertz: The derived SI unit for frequency; one hertz equals one cycle per second.

hickey: 1) A conduit bending tool. 2) A box fitting for hanging lighting fixtures. See Fig. H-4.

high-hat: A ceiling recessed incandescent lighting fixture of round cross-section, looking like a man's high hat in the shape of its construction. See Fig. H-5.

high-intensity discharge lamp (HID): Either a mercury, metal halide, and/or high-pressure sodium lamp.

high-leg: That phase conductor of a 3-phase, 4-wire, delta-connected system that is not connected to the single-phase power supply; the conductor with the highest voltage to ground; this phase conductor must be identified (per *NEC®*) and is commonly painted orange to provide such identification.

Fig. H-4: Pipe-bending tool for rigid steel conduit; normally called a "hickey."

high pressure cutout: Electrical control switch operated by the high side pressure that automatically opens an electrical circuit if too high head pressure or condensing pressure is reached.

high-pressure sodium lamp (HPS): A lamp that sends an electrical charge through sodium vapor at high pressures and temperatures to create a bright light.

high side: Parts of a refrigerating system that are under condensing or high side pressure.

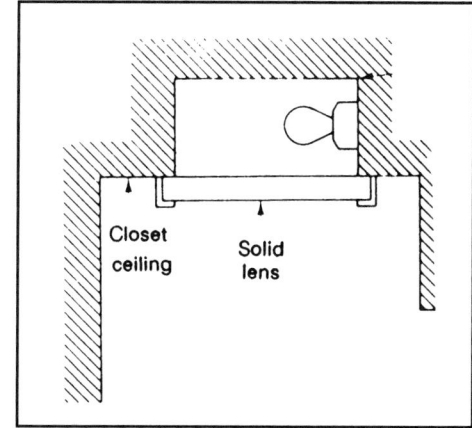

Fig. H-5: Recessed lighting fixture, sometimes called a "high-hat."

high-voltage control: All electrical controls above 5000 volts.

hi-pot test: A high-potential test in which equipment insulation is subjected to voltage level higher than that for which it is rated to find any weak spots or deficiencies in the insulation.

HMP, HMPE: High Molecular Weight Polyethylene.

holding contacts: Contacts that deliver current to a relay.

holding current: The minimum current to keep an SCR or a triac conducting.

hole: A mobile vacancy in the electron structure of semi-conductors that acts like a positive electron charge with mass.

Holtz motor: An ac motor that has a six-slotted rotor that can produce a speed of 1200 rpms.

homerun: That part of a branch circuit from the panelboard housing the branch circuit fuse or CB (overcurrent protection) and the first junction box at which the branch circuit is spliced to lighting or receptacle devices or to conductors that continue the branch circuit to the next outlet or junction box. The term "homerun" is usually reserved for multi-outlet lighting and appliance circuits. See Fig. H-6 on page 119.

horn: Alarm used when a fire system needs a loud audible system.

horsepower: The non-SI unit for power; 1 hp = 1 HP = 746 w (electric) = 9,800 w (boiler).

hot: Energized with electricity.

hot dip: Coating by dipping into a molten bath.

hot gas bypass: Piping system in a refrigerating unit that moves hot refrigerant gas from a condenser into the low pressure side.

hot junction: That part of the thermoelectric circuit that releases heat.

hot leg: A circuit conductor that normally operates at a voltage above ground; the phase wires or energized circuit wires other than a grounded neutral wire or grounded phase leg. The ungrounded conductor of a circuit.

Fig. H-6: Electrical floor plan showing outlets, circuits, and homeruns; the latter is indicated by half-arrowheads.

hot modulus: Stress at 100% elongation after 5 minutes of conditioning at a given temperature (normally 130°C).

hot stick: A long insulated stick having a hook at one end that is used to open energized switches, etc.

hot wire: A resistance wire in an electrical relay that expands when heated and contracts when cooled.

household fire alarm system: A fire alarm system designed for family dwellings that warns the occupants of smoke or fire.

housing: The enclosure that surrounds and protects the workings of a fixture, switch, receptacle, or appliances.

HTGR (Hi-Temp, Gas-Cooled Reactor) (nuclear power): A basic nuclear power fission reactor in which the reactor heats a gas; the gas exchanges its heat with a secondary loop to produce steam for the turbine or generator.

hub: 1) A fitting to attach threaded conduit to boxes. See Fig. H-7. 2) The central part of a cylinder into which a shaft may be inserted. 3) A reference point used for overhead line layout.

hum: Interference from ac power, normally of low frequency and audible.

Fig. H-7: Electric meter base with hub for conduit or WP SE cable connectors.

humidistat: A device that starts an exhaust fan when the humidity reaches a high level.

humidity: Moisture, dampness. Relative humidity is the ratio of the quantity of vapor present in the air to the greatest amount possible at a given temperature.

hybrid solid state relay: A relay that incorporates both electromechanical and solid state to solve unique problems.

hydraulic bender: A conduit bender that is equipped with a hydraulic hand pump or electric pump. This type of bender is very

Hypot® 121

Fig. H-8: Hydraulic pipe bender bending a 90° bend (elbow) in 4-inch rigid steel conduit.

useful when bending large sizes of conduit two inches and larger. See Fig. H-8.

hydraulic jack: A lifting jack, actuated by a small force pump enclosed within it, and operated by a lever from the outside.

hydromassage tub: Intended to be filled, used, then drained after each use. Also called a whirlpool tub.

hydrometer: Floating instrument used to measure specific gravity of a liquid. Specific gravity is the ratio of the density of a material to the density of a substance accepted as a standard. See Fig. H-9 on page 122.

hydronic: Type of heating system that circulates a heated fluid, usually water, through baseboard coils. Circulation pump is usually controlled by a thermostat.

hygrometer: An instrument used to measure the degree of moisture in the atmosphere.

hygroscopic: Readily absorbing and retaining moisture.

Hypalon®: The Dupont trade name for chlorosulfonated polyethylene.

Hypot®: Registered trade name of Associated Research, Inc. for their hi-pot tester.

Fig. H-9: A hydrometer is used to measure specific gravity of a liquid, especially in a storage battery.

hypotenuse: The side opposite the right angle of a triangle; it is the longest side of the triangle. See Fig. A-1.

hysteresis: The time lag exhibited by a body in reacting to changes in forces affecting it; an internal friction.

I

I^2r: The power losses in a device or apparatus due to resistance heating.

I^2t: Relating to the heating effect of a current (amps-squared) for a specified time (seconds), under specified conditions.

IACS: (International Annealed Copper Standard): Refined copper for electrical conductors; 100% conductivity at 20C for 1m × 1mm2 has $1/58$ ohm resistivity, 8.89 grams per mm2 density, 0.000 017 per degree C coefficient of linear expansion, $1/254.45$ per degree C coefficient of variation of resistance; however, NBS suggest the use of 8.93 for density (1977).

IAEI: International Association of Electrical Inspectors.

I beam: Rolled steel beam or built-up beam of an I section.

IBEW: International Brotherhood of Electrical Workers.

IC: Pronounced "eye see": 1) Interrupting capacity of any device required to break current (switch, circuit breaker, fuse, etc.); the amount of current that the device can interrupt without damage to itself. 2) Electronics: integrated circuit.

IC fixture: Recessed lighting that is rated for direct contact with insulation.

ID: Inside diameter.

IDCI: Immersion Detection Circuit Interrupter. Shuts off power to the device if dropped or put into water.

identified: Marked to be recognized as grounded.

identified conductor: Is the insulated grounded conductor that is white or light gray in color or is identified by three continuous white strips along its entire length for #6 or smaller. For conductors larger than #6 the identification is the same. A distinctive white mark at its terminations is also acceptable.

identified terminal: The grounded conductor terminal is silver or whitish in color. The hot conductor terminal is brass in color.

IEC: 1) Independent Electrical Contractors Inc., a national trade organization for independent electrical contractors with a BAT certified four-year apprenticeship program. 2) International Electromechanical Commission, an international standards organization that produces standards for electrical and electronic equipment. 3) International Electrochemical Commission.

IEEE: Institute of Electrical and Electronics Engineers.

IEGC: Isolated Equipment Grounding Conductor.

IESNA: Illuminating Engineering Society of North America.

ignition transformer: A transformer designed to provide a high voltage current.

IIR (Isobutylene Isoprene Rubber): Butyl synthetic rubber.

illuminance: The density of light on a surface.

impedance: 1) The opposition to current flow in an ac circuit; impedance includes resistance (R), capacitive reactance (X_C) and inductive reactance (X_L); it is measured in ohms. 2) The opposition to current flow in an ac transformer.

impedance matching: Matching source and load impedance for optimum energy transfer with minimum distortion.

impulse: A surge of unidirectional polarity.

incandescence: Glowing due to heat.

incandescent: That which gives light or glows at a white heat.

incandescent lamp: An electric lamp or bulb containing a thin wire or filament of infusible conducting material. See Fig. I-1.

Fig. I-1: Incandescent lamp.

inching: Momentary activation of machinery used for inspection or maintenance. Controls for such activation usually consists of *start*, *stop*, and *inch*. The device used to operate such machinery is usually the ac induction motor—either single- or three-phase.

incombustible material: Material that will not ignite or actively support combustion in a surrounding temperature of 1200 degrees Fahrenheit during an exposure of five minutes; also, material that will not melt when the temperature of the material is maintained at 900 degrees Fahrenheit for at least five minutes.

incoming wire: A hot wire that feeds power into a junction box.

indirect lighting: Lighting effect obtained by focusing the light against the ceiling or some other surface from which it is diffused in the area to be lighted. See Fig. I-2 on page 126.

indoor: Not suitable for exposure to the weather.

induced current: A magnetic field that can induce a current in a conductor when a wire passes through flux lines.

induced voltage: A voltage set up by a varying magnetic field linked with a wire, coil, or circuit.

inductance: The creation of a voltage due to a time-varying current; the opposition to current change, causing current changes to lag behind voltage changes; units of measure: henry.

Fig. I-2: Valances provide indirect up-light that reflects off the ceiling for general room lighting and down-light for drapery accent.

induction: The production of magnetization or electrification in a body by the mere proximity of magnetized or electrified bodies, or of an electric current in a conductor by the variation of the magnetic field in its vicinity.

induction coil: Essentially a transformer with open magnetic circuit, in which an alternating current of high voltage is induced in the secondary by means of a pulsating direct current in the primary.

induction heater: The heating of a conducting material in a varying electromagnetic field due to the material's internal losses.

induction machine: An asynchronous ac machine to change phase or frequency by converting energy—from electrical to mechanical, then from mechanical to electrical.

induction motor: An ac motor that does not run exactly in step with the alternations. Currents supplied are led through the stator coils only; the rotor is rotated by currents induced by the varying field set up by the stator coils.

inductive compensation: To improve the performance of a universal ac motor, a shorted compensating winding is used.

inductive kick: High counter-electromotive force introduced when an inductive circuit is opened caused by a collapsing magnetic field.

inductive reactance: The opposition to the flow of an electrical current in a circuit that consists of turns of wire. The opposition is greater if the turns are wound on an iron core. The measure of resistance to the flow of alternating current through a coil.

inductivity: The capacity or power for induction.

inductor: A device having winding(s) with or without a magnetic core for creating inductance in a circuit.

inertia: The property of matter that opposes any change in its state of motion.

infinite-position temperature control: Temperature control that allows any number of settings between zero and the maximum setting.

infrared lamp: An electrical device that emits infrared rays—invisible rays just beyond red in the visible spectrum.

infrared radiation: Radiant energy given off by heated bodies which transmits heat and will pass through glass.

inherently limited transformer: A transformer that is designed to limit the short circuit to a maximum of 8 amps.

ink: The material used for legends and color coding.

in phase: The condition existing when waves pass through their maximum and minimum values of like polarity at the same instant.

input percentage: The ratio of time that heating element contacts are in the closed position compared to the open position.

input status indicators: Potentiometers, barcode readers, and transducers are examples.

input voltage: The supply voltage to an electrical device or circuit.

inrush current: The current flow, immediately after the control device is turned on, to the load.

in sight: Within sight but not over 50 feet according to the *NEC*®.

in-socket switch: The socket of a lamp that has the switch included within it.

instantaneous trip breaker: A breaker that will trip in an instant when there is a short in the circuit.

instantaneous value: The value of a variable quantity at a given instant.

instrument: A device for measuring the value of the quantity under observation.

insulated: Separated from other conducting surfaces by a substance permanently offering a high resistance to the passage of energy through the substance.

insulating tape: Adhesive tape made nonconducting by being saturated with an insulating compound or manufactured from a nonconducting material; used for covering wire joints and exposed parts.

insulating transformer: A transformer that has the primary carefully insulated from the secondary. There is no electrical metallic connection between the primary and secondary.

insulation, class rating: A temperature rating descriptive of classes of insulations for which various tests are made to distinguish the materials; not related necessarily to operating temperatures.

insulation, electrical: A medium in which it is possible to maintain an electrical field with little supply of energy from additional sources; the energy required to produce the electric field is fully recoverable only in a complete vacuum (the ideal dielectric) when the field or applied voltage is removed: used to a) save space; b) enhance safety; c) improve appearance.

insulation fall-in: The filling of strand interstices, especially the inner interstices, which may contribute to connection failures.

insulation level (cable): The thickness of insulation for circuits having ground fault detectors which interrupt fault current within: 1) 1 minute = 100% level; 2) 1 hour = 133% level; 3) over 1 hour = 173% level.

insulation, thermal: Substance used to retard or slow the flow of heat through a wall or partition.

insulator: Any material that will not allow current to flow.

integral: Built into or self-contained; that is, an electric heater with an integral thermostat.

integrated circuit: A circuit in which different types of devices such as resistors, capacitors, and transistors are made from a single piece of material and then connected to form a circuit.

integrated electrical system: An industrial wiring system in which:

- An orderly shutdown is required to minimize personnel hazard and equipment damage. The conditions of maintenance and supervision assure that qualified persons will service the system.
- Effective safeguards, acceptable to the authority having jurisdiction, are established and maintained.

integrator: Any device producing an output proportionate to the integral of one variable with respect to a second variable; the second is usually time.

intercalated tapes: Two or more tapes of different materials helically wound and overlapping on a cable to separate the materials.

interchangeable wiring devices: Single switches, pilot lights, and receptacle outlets that can be mounted and installed in an outlet box. Up to three devices can be installed in the same space as a single duplex receptacle.

interconnected system: Operating with two or more power systems connected through tie lines.

interface: 1) A shared boundary. 2) (nuclear power); a junction between Class 1E and other equipment of systems. 3) The point where two different systems that have different characteristics interact and exchange information, such as keyboards, touch screens, and transducers.

interference: Extraneous signals or power that are undesired.

interlock: A safety device to ensure that a piece of apparatus will not operate until certain conditions have been satisfied.

internal clamp: A device that screws to the inside of a junction box to hold electrical cables in place.

internal relay: A digital programmable controller that is capable of being programmed to function like a control relay.

internal resistance: The natural resistance that is contained within a power source.

interpolate: To estimate an intermediate between two values in a sequence.

interpole: A small field pole placed between the main field poles and electrically connected in series with the armature of an electric rotating machine. See Fig. I-3.

interrupter: A device that opens and closes a circuit at very frequent intervals.

interrupting rating: The highest current or voltage a device is designed to interrupt under specified conditions.

interrupting time: The sum of the opening time and arcing time of a circuit opening device.

Fig. I-3: Interpoles are the small poles placed between the field poles and connected in series.

interstice: The space or void between assembled conductors and within the overall circumference of the assembly.

intrinsically safe: Incapable of releasing sufficient electrical or thermal energy under normal or abnormal conditions to cause ignition of a specific hazardous atmospheric mixture in its most ignitable concentration.

intumescence: The capability of a material to expand when heated.

inverse time relay: The characteristic of this relay is to prevent tripping on normal starting current and on harmless momentary overloads.

inverter: An item that changes dc to ac. See Fig. I-4 on page 131.

ion: An electrically charged atom or radical.

ionization: The process or the result of any process by which a neutral atom or molecule acquires charge. A breakdown that occurs in gaseous parts of an insulation when the dielectric stress exceeds a critical value without initiating a complete breakdown of the insulation system; ionization is harmful to living tissue, and is detectable and measurable; may be evidenced by corona.

ionization smoke detector: Contains low-level radioactive particles that ionize the air in the detector's sensing chamber, causing a small current to flow that sets off the alarm. See Fig. I-5 on page 132.

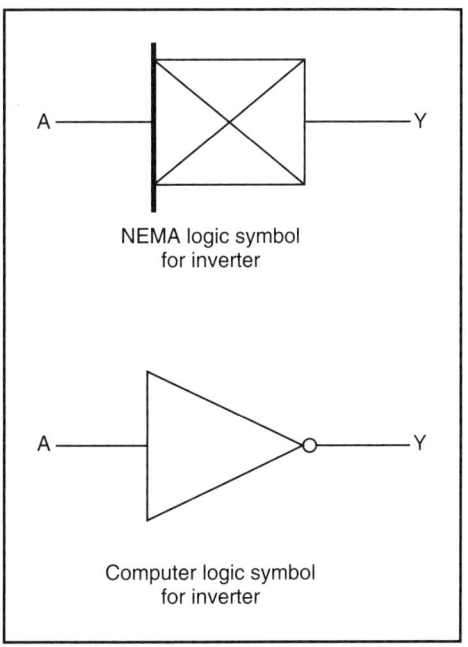

Fig. I-4: Inverter symbols for NEMA and computer logic.

ionization factor: The difference between percent dissipation factors at two specified values of electrical stress; the lower of the two stresses is usually so selected that the effect of the ionization on dissipation factor at this stress is negligible.

IPCEA (Insulated Power Cable Engineers Association): The association of cable manufacturing engineers who make nationally recognized specifications and tests for cables.

IR (Insulation Resistance): The measurement of the dc resistance of insulating material; can be either volume or surface resistivity; extremely temperature sensitive.

IR drop: The voltage drop across a resistance due to the flow of current through the resistor.

IRK (Insulation dc Resistance Constant): A system to classify materials according to their resistance on a 1000 foot basis at 15.5°C (60°F).

Fig. I-5: Characteristics of an ionization smoke detector.

iron-vane meter: Used to measure low frequency ac.

irradiation, atomic: Bombardment with a variety of subatomic particles; usually causes changes in physical properties.

ISO: International Organization for Standardization, which has put together the "SI" units that are now the international standards for measuring.

isolated: Not readily accessible to persons unless special means for access are used.

isolated ground receptacle: The green grounding hexagon screw that connects to a grounding conductor and does not tie to any ground references in between. See Fig. I-6.

isolating: Referring to switches, this means that the switch is not a loadbreak type and must only be opened when no current is flowing in the circuit. This term also refers to transformers (an isolating transformer) used to provide magnetic isolation of one circuit from another, thereby breaking a metallic conductive path between the circuits.

Fig. I-6: Isolated ground receptacle is a separate ground to the receptacle.

isolation transformer: A transformer whose secondary winding is electrically isolated from the primary winding.

isometric drawing: A method of drawing that shows three sides of an object. All vertical lines are drawn vertically and the horizontal lines are on a 30° plane.

isotope: Atoms of a given element, each having different mass from the other because of different quantities of subatomic particles in the nucleus; isotopes are useful because several are naturally radiating and thus can become radiation sources for medical treatments or researching labs; a common isotope is Cobalt 60.

J

jack: A plug-in type terminal. See Fig. J-1.

jacket: A non-metallic polymeric close-fitting protective covering over cable insulation; the cable may have one or more conductors.

jacket, conducting: An electrically conducting polymeric covering over an insulation.

jamb: Upright member forming the side of a door or window opening.

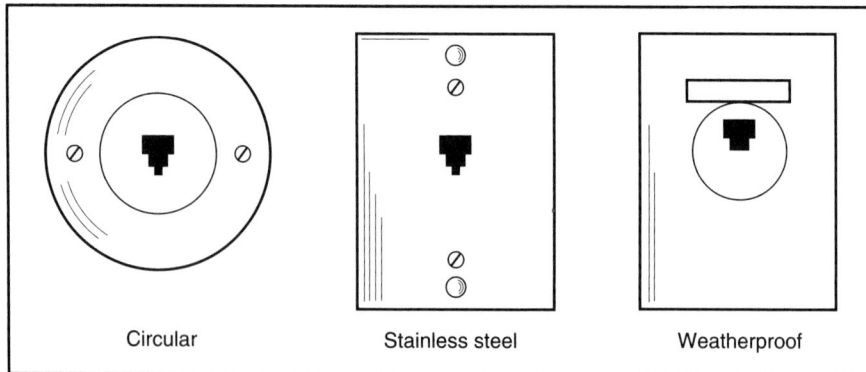

Fig. J-1: Different types of telephone jacks.

jamming: The wedging of a cable so that it can no longer be moved during installation.

jan: Joint army and navy specification.

JB: Pronounced "jay bee"; refers to any junction box, taken from the initial letters J and B.

J-block: A siding term; when installed in siding allows installation of electrical box to set flush.

jigsaw: A small hand reciprocating saw that is capable of cutting scroll work. Used to cut metal, wood, plastics, and sheetrock.

jog station: One or more control devices added to the control circuit, which allows the motor to run as long as the jogging button is engaged.

joists: Horizontal members designed to carry the load of a floor or ceiling.

joule: The derived SI unit for energy, work, quantity of heat; one joule equals one newton-meter or .73732 foot pounds. One joule per second equals one watt.

joule heat: Also called *joule effect*, the thermal effect that results when electrical current flows through a resistance. It is measured in watts. When a current of 1 A flows through a resistance of 1 Ω, the joule heat given off is equivalent to 1 W. This is explained by Ohm's law for power, which reads:

$$P = I^2 R$$

where P is power in watts, I is current in amperes, and R is resistance in ohms. Using this formula, when 2 A of current flow through a resistance of 1 Ω, the total power dissipation is 4 W.

Joule's law: The heat generated in a conductor by an electric current is proportional to the resistance of the conductor, the time during which the current flows, and the square of the strength of the current. The quantity of heat in calories may be calculated by the use of the equation:

$$\text{Calories per second} = \text{Volts} \times \text{Ampere} \times 0.24$$

The total number of calories or heat developed in seconds will be given by:

$$\text{Heat} = \text{Volts} \times \text{Amperes} \times \text{Seconds} \times 0.24$$

journeyman: Properly, one who has gained a thorough knowledge of his or her trade by serving an apprenticeship, although the term is often applied to any worker who is sufficiently skilled to command the standard rate of journeyman's pay.

jumper: A short length of conductor, usually a temporary connection. See Fig. J-2.

junction: A connection of two or more conductors. Place of union; point of meeting; joint.

junction box: An enclosure where one or more raceways or cables enter, and in which electrical conductors can be, or are, spliced. See Fig. J-3.

Fig. J-2: Bonding jumper wires used to ensure continuity at a water meter. See *NEC*®

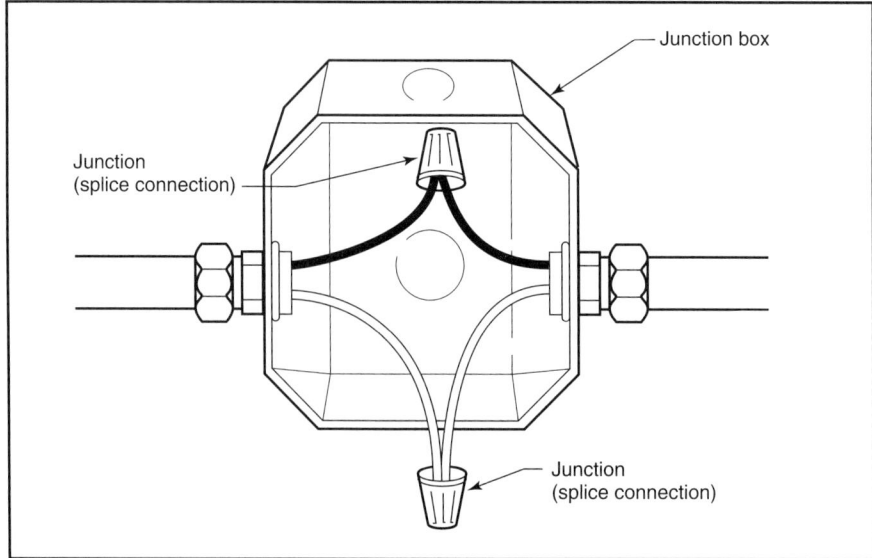

Fig. J-3: Junction box is a box in which one or more wires enter and are spliced.

junction diode: P-type material and N-type material that are used on the same component to form a junction, creating an area sensitive to voltage changes.

jurisdiction: The area assigned to a local labor union.

jute: A fibrous natural material used as a cable filler or bedding.

ka: KiloAmpere.

kc: Kilocycle; use kiloHertz.

kcmil: One thousand circular mils. 250 kcmil is a conductor size of 250,000 circular mils; 500 kcmil is 500,000 circular mils, etc.

kel: Kilogram.

Kelvin (K): The basic SI unit of temperature: $1/273.16$ of thermodynamic temperature of the triple freezing point of water.

Kelvin double bridge: A special bridge that is used for measuring very low resistance (0.1 ohm or less). The arrangement of the bridge reduces the effects of contact resistance, which causes significant error when such low resistances are connected to conventional resistance bridges.

Kelvin's temperature: Any temperature in the absolute scale. This is a temperature scale with its zero point at –273.1°C, or absolute zero. The unit of thermodynamic temperature is the *kelvin*, and its symbol is K.

keyhole saw: A small tapered-blade saw used for cutting keyholes, small openings for fishing wires, and the like. See Fig. K-1 on page 139.

kHz: kiloHertz. A unit of frequency that is equivalent to 1000 Hz, or 1000 cycles.

kick-back diode: Used to stop inductive spike voltages produced in the armature of a motor.

kilo: 10^3 or 1000.

Fig. K-1: Keyhole saw.

kilocycle: One thousand cycles.

kilogram (kg): The basic SI unit for mass; an arbitrary unit represented by an artifact kept in Paris, France.

kilometer: A metric unit of linear measurement equal to 1000 meters.

kilovolt ampere: One thousand volt-amperes.

kilowatt: Unit of electrical power equal to 1000 watts.

kilowatt-ft: The product of load in kilowatts and the circuit's distance over which a load is carried in feet; used to compute voltage drop.

kilowatt hour: The work performed by one kilowatt of electric power in one hour. The unit on which the price of electrical energy is based.

kinetic energy: Energy by virtue of motion. Whenever work is accomplished on an object, energy is consumed (changed from one kind to another). If no energy is available, no work can be performed. Thus, energy is the ability to do work. One form of energy is that which is contained by an object in motion. In driving a nail into a block of wood, a hammer is set in motion in the direction of the nail. As the hammer strikes the nail, the energy, or motion of the hammer, is converted into work as the nail is driven into the wood. This energy is called *kinetic energy*.

Kirchhoff's Laws: 1) The algebraic sum of the currents at any point in a circuit is zero. That is:

$$I_1 + I_2 + I_3 + \ldots = 0$$

where I_1, I_2, I_3, etc. are the currents entering and leaving the junction. Currents entering the junction are assumed to be positive, while currents leaving the junction are negative. When solving a problem using this equation, the currents must be placed into the equation with the proper polarity signs attached. 2) The algebraic sum of the product of the current and the impedance in each conductor in a circuit is equal to the electromotive force in the circuit.

knee: The bend in a response curve that is most often an indication of the onset of saturation or cutoff. The point on the curve that represents the knee marks an abrupt change.

knife-blade fuse: A fuse having end connections that resemble the blades of a knife switch and that fit into the contacts of the cutout in the same manner that the blades of a switch fit into the switch contacts.

knife switch: A switch that opens or closes a circuit by the contact of one or more blades between two or more flat surfaces or contact blades. See Fig. K-2.

knob: A porcelain device for holding electrical conductors in place. For open wiring on insulators, the knob is used in conjunction with cleats, tubes, and flexible tubing for the protection and support of single insulated conductors run in or on buildings, and not concealed by the building structure. When concealed, this wiring method is called "concealed knob-and-tube" wiring. Open wiring on insulators has been limited to farm and industrial applications; concealed knob-and-tube wiring is allowed only as an extension of an existing system or else with special permission from the inspecting authorities. See Fig. K-3 on page 141.

Fig. K-2: Knife switch.

knockout: A portion of an enclosure designed to be readily removed for installation of a raceway or cable connector. Sometimes called "concentric" or "eccentric." See Fig. K-4 on page 141.

knock-out punch: A tool that cuts round holes in metal panel boxes, distribution boxes, etc. The cut holes are sized to fit the electrical device that is to be connected to the panel or distribution box. See Fig. K-5 on page 142.

Fig. K-3: Open wiring on insulators is allowed only for industrial and farm applications.

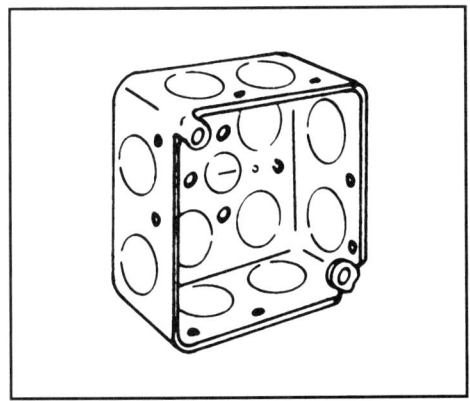

Fig. K-4: An outlet box showing knockouts.

Fig. K-5: A knockout punch and case are used to cut holes in metal boxes.

KO: Pronounced "kay oh"; a knockout, the partially cut opening in boxes, panel cabinets and other enclosures, which can be easily knocked out with a screw driver and hammer to provide a clean hole for connecting conduit, cable or some fittings. See *knockout*.

kVA: Kilovolts times Ampere. Also referenced as 1000 volt-amperes.

L

LA: Lightning Arrestor.

labeled: Items to which a label, trademark, or other identifying mark of nationally recognized testing labs has been attached to identify the items as having been tested and meeting appropriate standards.

lacquer: A protective coating or finish that dries to form a film by evaporation of a volatile constituent.

lagging: 1) The wood covering for a reel. 2) The time relationship between voltage and current in an ac circuit when the voltage reaches its maximum after the current reaches its maximum. The voltage is lagging the current.

lally column: Concrete-filled cylindrical steel structural column.

laminated core: An assembly of steel sheets for use as an element of magnetic circuit; the assembly has the property of reducing eddy-current losses.

laminated wood: Wood built up of piles or laminations that have been joined either with glue or with mechanical fasteners. The piles usually are too thick to be classified as veneer, and the grain of all piles is parallel.

Fig. L-1: Several types of lamps in current use.

lamp: A device to convert electrical energy to radiant energy, normally visible light; usually only 10–20% is converted to light. See Fig. L-1.

lamp temperature: The exterior temperature of a lamp when it is turned on.

LAN: Local Area Network.

lap: The relative position of applied tape edges; "closed butt lap"—tapes just touching; "open butt" or "negative lap"—tapes not touching; "positive lap" or "lap"—tapes overlapping.

lap winding: A parallel winding that has an even number of segments and slots. The number of bars, however, is equal to or a multiple of the number of slots. It is known as a parallel winding because there are as many circuits in parallel as there are poles. Thus, a four-pole motor has four circuits in parallel, and a six-pole motor has six circuits in parallel through the armature. This requires that there be as many brushes as there are poles for the motor.

Since there are so many paths in parallel, a larger current can pass through the winding because it has several paths to travel. This, in turn, reduces the amount of voltage needed in the armature to push the current through so many paths.

The lead connections for a lap winding come from each side of the coil as top and bottom leads and connect adjacent to each other on the commutator. The position that they connect on the commutator cannot be set down to a hard-and-fast theoretical rule, but is rather at the discretion of the manufacturer who designs the motor.

laser: A device that emits visible radiation.

laser diode: Diode that has an optical cavity that has very high reflective surfaces. Used in transmitters for single-mode fiber links.

latent heat: Heat given off or absorbed in a process (as vaporization or fusion) other than a change in temperature.

latitude adjustment head: A device that allows a satellite dish to be tilted in various directions.

lattice structure: The name of the pattern when semiconductors are refined to their pure form.

law of charges: Like charges repel, unlike charges attract.

law of magnetism: Like poles repel, unlike poles attract.

lay: The axial length of one turn of the helix of any element in a cable.

lay direction: Direction of helical lay when viewed from the end of the cable.

lay-in-fixture: Recessed fluorescent fixture designed to lay in the grid work framing of suspended ceilings. See Fig. L-2 on page 146.

lay length: Distance along the axis for one turn of a helical element.

LB fitting: A 90° conduit connector that has an access port to bring cable through a limited space. See Fig. L-3 on page 146.

lead (leed): A short connecting wire brought out from a device or apparatus.

Fig. L-2: Lay-in fixture designed to fit in suspended ceilings.

lead-acid battery: The most widely used type of storage battery, which has an EMF (voltage) of 2.2 V per cell. In its charged condition, the active materials in the lead-acid battery are lead dioxide (sometimes referred to as lead peroxide) and spongy lead. The lead dioxide is used as the positive plate, while the spongy lead forms the negative plate.

leading: 1) Applying a lead sheath. 2) When volt-

Fig. L-3: LB fitting is a 90° conduit connector that has an access port to bring cable through a wall.

age reaches its maximum amplitude earlier than the current, the voltage is leading the current.

lead squeeze: The amount of compression of a cable by a lead sheath.

leakage: Undesirable conduction of current.

leakage distance: The shortest distance along an insulation surface between conductors.

leakage resistance: The ohmic value of the path between two electrodes that are insulated from each other.

LED: See *light-emitting diode*.

left-hand generator rule: With the thumb, forefinger, and the middle finger of the left hand perpendicular to each other, point the forefinger in the direction of the flux and the thumb in direction of the conductor. The middle finger is now pointing in the direction in which the generator tends to send current through the circuit. See Fig. L-4.

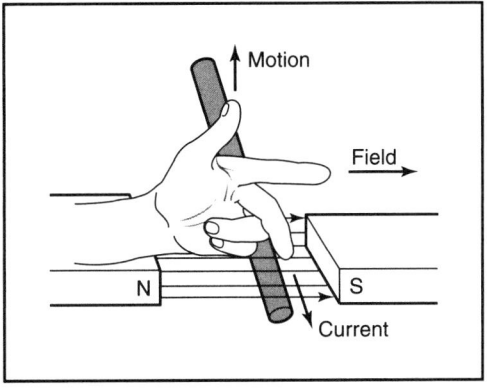

Fig. L-4: Left-hand generator rule. The forefinger points in the direction of flux. The thumb points in the direction of conduction. The middle finger points in the direction of current flow.

left-hand rule: When the left-hand fingers wrap around the wire with the thumb pointing in the direction of the current flow, the fingers will be pointing in the direction of the magnetic field. See Fig. L-5 on page 148.

leg: A portion of a circuit, such as a switch leg or switch loop.

legend, embossed: Molded letters and numbers in the jacket surface; letters may be raised or embedded.

legend plate: A plate mounted on a switch showing the switch's function.

leg length (pipe bending): The distance from the center line of a bend to the end of the straight section of pipe.

Lenz' Law: "In all cases the induced current is in such a direction as to oppose the motion which generates it." That which states that whenever the value of an electric current is changed in a circuit, it creates an electromotive force by virtue of the energy stored up in its magnetic field, which opposes the change.

light: Radiant energy lying within a wavelength that spreads from 100 to 10,000 nm. The human eye can perceive light radiations in a frequency range of between 450 and 700 nm. The color of light is determined by its wavelength. Energy at the short-wave end of the visible spectrum, from 380 to about 450 nm, produces the sensation of violet. The longest visible waves, from approximately 630 to 760 nm, appear as red. Between these lie the wavelengths that the eye sees as blue (450–490 nm), green (490–560 nm), yellow (560–590 nm), and orange (590–630 nm), the colors of the rainbow.

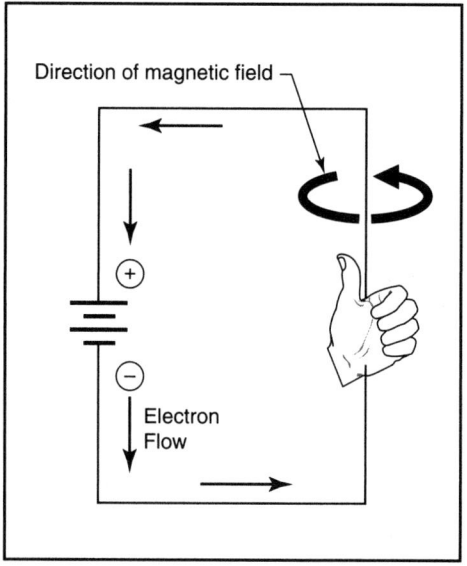

Fig. L-5: Left-hand rule is used to determine the direction of the magnetic field.

light-emitting diode (LED): A semiconductor diode that produces a light when the current is flowing through it.

lighting and appliance panelboard: A panelboard that has more than 10% of the overcurrent devices rated at 30 amps or less.

lighting outlet: An outlet installed solely for the purpose of installing some type of lighting device.

lightning: A natural static electrical discharge that goes from cloud to cloud, ground to cloud, or cloud to ground.

lightning arrestor: A device designed to protect circuits and apparatus from high transient voltage by diverting the over-voltage to ground. They should be placed on upward projections such as chimneys, towers, and the like. On flat roofs, rods should be placed

50 ft on center, and on edges of flat roofs and ridges of pitched roofs, about 25 ft on center. The rods should project from 10 to 60 in. above flat roofs and ridges, and from 10 to 14 in. above upward projections. See *surge arrestor.*

lightning rod: A device that is mounted on a roof to accept the lightning charge and send the charge directly to ground.

light quantity: A measure of the total light emitted by a source of light falling on a surface. It may be expressed in lumens or watts. If all the light were of green-yellow color, at the peak of the spectral luminosity curve, 1-W would be equal to 681 lm. This gives some idea of the relative luminous efficiency of common light sources. For instance, a 100-W incandescent lamp emits about 1600 lm, only 2.4 W as light and the balance as heat. A 40-W fluorescent lamp emits about 3100 lm, 4.5 W as light and the balance as heat. From this, it can be seen that the fluorescent lamp gives four to five times as much light as the incandescent on a watt-for-watt basis.

light relay: A photoelectric device that triggers a relay in accordance with fluctuations in the intensity of a light beam. These may be purely electronic in nature or electromechanical, depending on the individual circuit. Often, light relays use an active or passive photoelectric device in the base circuit of a transistor, which is connected in series with an electromechanical relay. When light strikes the photocell, the transistor is driven to saturation and conducts current through the relay coil, causing it to change states.

light wave: A stream of electromagnetic energy that falls into the light spectrum. This lies between 100 and 10,000 nm. Visible light (that which can be seen by the human eye) lies within a very narrow band width of this spectrum at a frequency range of about 500 to 600 nm.

LIM (Laboratory Inspection Manual): A summary of specified values for quality testing.

limit: A boundary of a controlled variable.

limit control: Control used to open or close electrical circuits as temperature of pressure limits are reached.

limiter: A device in which some characteristic of the output is automatically prevented from exceeding a predetermined value.

line: A circuit between two points; ropes used during overhead construction.

linear: Arranged in a line.

linearity: When the effect is directly proportional to the cause.

line, bull: A rope for large loads.

line current: Same as phase current in a wye-connect system.

line diagram (ladder diagram): The control language that shows the logic of an electrical circuit and its electrical components. Does not show physical location, as does a wiring diagram. See Fig. L-6.

Fig. L-6: Line or ladder diagram.

line drop: The amount of voltage drop is directly related to the amount of resistance in a conductor or equipment.

line, finger: A rope attached to a device on a pole when a device is hung, so further conductor installation can be done from the ground.

lineman's pliers: A pair of pliers that is designed for both gripping and cutting is used by electricians and linemen in construction work. See Fig. L-7.

line, pilot: A small rope strung first.

line side: The supply side of any equipment or device.

lines of flux: The lines of force in a magnetic field.

line, tag: A rope to guide devices being hoisted.

Fig. L-7: Lineman's pliers are used for gripping and cutting cable.

line-to-line voltage: The voltage measured between the lines of a single line winding in a wye-connection.

liquid absorbent: A chemical in liquid form that has the property to absorb moisture.

liquid immersion control: A control that can be immersed in liquid and that uses its sensors to control levels and temperatures of the liquids.

liquid line: The tube that carries liquid refrigerant from the condenser or liquid receiver to the refrigerant control mechanism.

liquid receiver: Cylinder connected to a condenser outlet for storage of liquid refrigerant in a system.

Lissajous Figure: A special case of an x-y plot in which the signals applied to both axes are sinusoidal functions; useful for determining phase and harmonic relationships.

listed: Items in a list published by a nationally recognized independent lab that makes periodic tests. See Fig. L-8, on page 152. See *labeled*.

liter: Metric unit of volume that equals 61.03 cubic inches.

live: Energized.

live-front: Any panel or other switching and protection assembly, such as switchboard or motor control center, which has exposed electrically energized parts on its front, presenting the possibility of contact by personnel.

live load: Any load on a structure other than a dead load includes the weight of persons occupying the building and freestanding material.

Fig. L-8: Underwriters' Laboratories' label.

LMFBR (Liquid Metal, Fast-Breeder Reactor) (nuclear power): A basic nuclear power fission reactor in which a metal (sodium) is heated by the reactor; this heats a second metal loop that transfers its heat to a third loop to produce steam for the turbine; this reactor can also produce other nuclear fuel at the same time.

LMP: Low Molecular Weight Polyethylene.

load: 1) A device that receives power. 2) The power delivered to such a device.

load-break: Referring to switches or other control devices, this phrase means that the device is capable of safely interrupting load current—to distinguish such devices from other disconnect devices that are not rated for breaking load current and must be opened only after the load current has been broken by some other switching device.

load calculations: The calculation of computed electrical loads.

load center: 1) An assembly of circuit breakers or switches. 2) The point where the branch circuit's power originates. See Fig. L-9, on page 153.

load factor: The ratio of the average to the peak load over a certain period. The time may be either the normal number of operating hours per day or 24 h, as generally used by the power companies.

Local fire alarm systems

Fig. L-9: A safety switch acting as a disconnect for a load center or panelboard.

The average load is equal to the kilowatt-hours used in the specified time, as measured by a watthour meter, divided by the number of hours. The maximum load is the highest load at one time, as measured by some form of maximum-demand or curve-drawing watthour meter.

load losses: Those losses incidental to providing power.

load side: Any current used after the overcurrent protection.

LOCA (Loss of Coolant Accident) (nuclear power): The test to simulate nuclear reactor accident exhibited by high radiation, high temperature, etc.

local building codes: The rules and regulations that are set by the local governing body that regulates the local building codes.

local control: A control function or an initiation or a change that is accomplished at the same place as the electric controller.

local fire alarm systems: Now called "protective premises fire alarm systems." In the NFPA 72, the definition of protected premises is "the physical location protected by the alarm system."

local lighting: Lighting that is focused on a certain area.

location, damp: A location subject to a moderate amount of moisture such as some basements, barns, cold-storage, warehouses, and the like.

location, dry: A location not normally subject to dampness or wetness; a location classified as dry may be temporarily subject to dampness or wetness, as in the case of a building under construction.

location, wet: A location subject to saturation with water or other liquids.

locked rotor: The condition occurring when the circuits of a motor are energized but the rotor is not turning.

locked rotor torque: The same as starting torque; it is the torque needed to start a motor.

locking switch: A switch that must be unlocked before it can be put in the "on" position.

locknut: A fitting for securing cable and conduit connectors to outlet boxes, panelboards, and other apparatus; also used on threaded metal conduit. See Fig. L-10.

lockout: To keep a circuit locked open. To "tag out" a circuit as per OSHA rules.

lock seam: Joining of two sheets of metal consisting of a folded, pressed, and soldered joint.

logarithm: The exponent that indicates the power to which a number is raised to produce a given number.

Fig. L-10: Locknut.

logarithmic: Pertaining to the function y = log x.

logic: The basic functions are: AND, OR, AND/OR, NOT, NOR, and NAND. These provide the basic steps to solve complex problems through the use of basic functions.

longwall machine (mining): A machine used to undercut coal at relatively long working spaces.

looping-in: Avoiding splices by looping wire through device connections instead of cutting the wire.

loss: Power expended without doing useful work.

low noise amplifier: Amplifies the satellite signal when received.

low pressure sodium lamp: A lamp that helps light highways, parking lots, bridges, etc. It gives off a yellowish light.

low voltage protection: Takes motors out of the circuit and does not allow them to be automatically restarted when loss of full power occurs.

low voltage release (LVR): A device that allows motors to be automatically restarted when full power has been restored.

lug: A device for terminating a conductor to facilitate the mechanical connection. Fig. L-11, on page 156.

lumen: The derived SI unit for luminous flux.

luminaire: A complete lighting unit that includes the lamp, sockets, and equipment for controlling light, such as reflectors and diffusers. On electric discharge lighting, the luminaire also includes a ballast. The common term used for luminaire is *lighting fixture*, or in some cases, simply *fixture*. See Fig. L-12, page 157.

luminance: Intensity in a given direction divided by projected area, as intercepted by the air. It is subjective intensity and ranges from very dim to very bright. Luminance is expressed as candelas per square inch in a certain direction. Candelas per square inch may be put into more convenient form by multiplying by 452, giving luminance in footlamberts. Another way of looking at luminance is in relation to illumination and the reflection factor. For a nonspecular surface:

$$\text{Luminance} = \text{Illumination} \times \text{Reflection factor}$$

or

$$L = E \times R$$

where E equals footcandles, R equals reflection factor, and L equals footlamberts.

Fig. L-11: Several types of lugs used for wire connections.

Fig. L-12: Several types of incandescent luminaires (lighting fixtures).

To illustrate, if $E = 100$ fc and $R = 50\%$, then $L = 100 \times 0.50 = 50$ footlamberts.

lus: The derived SI unit for illuminance.

lux: The international unit for illuminance.

LV: Low voltage.

LWBR (nuclear): Light Water Breeder Reactor.

M

machine: An item to transmit and modify force or motion, normally to do work.

machine language: A controlling language that is written in the binary format.

machine rating: The amount of power a machine can deliver without overheating.

machine screw: A very commonly used type of screw with clear-cut threads and a variety of head shapes. It may be used either with or without a nut.

Madison HOLDITS: The brand name of metal supports used in remodeling or old work to tie a recessed metal box to a ceiling or wall.

magnet: A body that produces a magnetic field external to itself; magnets attract iron particles.

magnet brake: Also called electric braking when dc voltage is applied to the stationary windings immediately after ac voltage is removed from the motor circuit, which causes a magnetic field that does not change polarity. This action brings the motor to a stop smoothly and quickly.

magnetic circuit breaker: An electromagnetic device for opening a circuit.

magnetic coil: The winding of an electromagnet. A coil of wire wound in one direction, producing a dense magnetic field capable of attracting iron or steel when carrying a current of electricity.

magnetic contactor: An electromagnetically operated switch.

magnetic controller: A device whose functions are controlled by electromagnets.

magnetic cutout: A device for breaking an electrical circuit by means of an electromagnet, instead of by fusing a part of the circuit.

magnetic density: The number of lines of force or induction per unit area taken perpendicular to the induction. In free space, flux density and field intensity are the same numerically, but within magnetic material, the two are quite different.

magnetic domain: The molecules in certain metals (for example, iron, nickel, and cobalt) act like small magnets.

magnetic field: 1) A magnetic field is said to exist at a point if a force over and above any electrostatic force is exerted on a moving charge at the point. 2) The force field established by ac through a conductor, especially a coiled conductor.

magnetic flux: Magnetic lines of force set up by an electromagnet, permanent magnet, or solenoid.

magnetic force: The force by which attraction and repulsion are exerted by the poles of a magnet.

magnetic induction: The number of magnetic lines or the magnetic flux per unit of cross-sectional area perpendicular to the direction of the flux.

magnetic permeability: A measure of the ease with which magnetism passes through a substance.

magnetic pole: Those portions of the magnet toward which the external magnetic induction appears to converge (south) or diverge (north).

magnetic switch: A switch operated or controlled by an electromagnet.

magnetism: 1) That property of iron, steel, and some other substances, by virtue of which, according to fixed laws, they exert forces of attraction and repulsion. 2) The science that treats the conditions and laws of magnetic force.

magnetoelectric: Electricity that is produced by a magnetic force flowing through a conductor or coil.

magnetomotive force (mmf): The effort exerted that creates a magnetic field. See *motors*.

main: The primary circuit that supplies all other circuits.

make and break: 1) The term may be applied to several electrical devices. Primarily, there is a pair of contact points, one stationary, and the other operated by a cam that makes the break in a circuit between these points. 2) Make closes the circuit, break opens the circuit.

Malibu light: Low voltage lighting intended to soft light an area such as a swimming pool vicinity.

mandatory rules: Pertains to the *NEC®*; when *shall* or *shall not* are used, the rules are mandatory.

manhole: A subsurface chamber, large enough for a man, to facilitate cable installation, splices, etc. in a duct bank.

manual: 1) Operated by mechanical force applied directly by personal intervention. 2) A handbook of instructions.

manual time-delay switch: A switch whose timing mechanism is set manually.

marker: A tape or colored thread in a cable that identifies the cable manufacturer.

masonry anchors: An expandable type of anchor that expands when the screw is inserted. A special drill bit is needed to form the hole for the anchor to set in.

mass: The property that determines the acceleration the body will have when acted upon by a given force; unit = kilogram.

master coded system: A fire alarm system that sends the same coded signal from all the fire alarms in the building.

master switch: A main switch; a switch controlling the operation of other switches.

mast-type service: The electrical service is above ground and runs from a utility pole to a service head that is mounted on a building.

mat: A concrete base for heavy electrical apparatus, such as transformers, motors, generators, etc.; sometimes the term includes the concrete base, a bed of crushed stone around it and an enclosing chain-link fence.

matrix: A multi-dimensional array of items.

matter: Any substance that occupies space.

matte surface: A surface from which reflection is predominately diffused.

maximum continuous current (mcc): A value that is determined by the manufacturers of hermetic AC compressors. *NEC*® states that the current can be no higher than 156% of the marked rated amps.

maximum overcurrent protection: A device that senses the current in a circuit and opens the circuit when current exceeds a preset level.

maximum voltage: The highest voltage reached in each alternation of an alternating voltage.

MCM: An expression referring to conductors of sizes from 250 MCM, which stands for Thousand Circular Mils, up to 2000 MCM. The newest way of expressing thousand circular mils is kcmil; that is, 250 kcmil, 500 kcmil, etc.

mean: An intermediate value; arithmetic—sum of values divided by the quantity of the values; the average.

mechanical advantage (rigging): Using multiple blocks increases the lifting capacity. See Fig. M-1 on page 163.

mechanical water absorption: A check of how much water will be absorbed by material in warm water for seven days (mg/sq. in. surface).

medium base: The regular base that a normal incandescent light uses. One of the most common is the AKA Edison base. One size smaller is called the candelabra base and one size larger is called the mogul base.

Fig. M-1: Mechanical advantage. Multiple blocks (B) require half the pull as a single block (A).

medium hard: A relative measure of conductor temper.

medium voltage control: Controls and apparatus that are rated at 600 to 5000 volts.

meg or mega: When prefixed to a unit of measurement it means one million times that unit and is abbreviated M.

megahertz (MHZ): A million cycles per second.

megavolt: A unit of voltage equal to one million volts.

megger®: A test instrument for measuring the insulation resistance of conductors and other electrical equipment; specifically, a megohm (million ohms) meter; this is a registered trade name of the James Biddle Co. See Fig. M-2, on page 164.

megohm: A unit of electrical resistance equal to one million ohms.

megohmmeter: An instrument for measuring extremely high resistance.

melt index: The extrusion rate of a material through a specified orifice at specified conditions.

melting time: That time required for an overcurrent to sever a fuse.

mercury vapor lamp: A type of lamp that produces light by passing a current through mercury vapor.

mesh: The simplest form of a dc loop.

Fig. M-2: Megger manufactured by the James Biddle Co.

mesh current: A type of circuit analysis. Also called pane mesh analysis. It is used to set up simultaneous equations in order to resolve unknown variables. Useful when there is more than one power source in a circuit.

messenger: The supporting member of an aerial cable.

metal clad (MC): The cable core is enclosed in a flexible metal covering.

metal-clad switchgear: Switch-gear having each power circuit device in its own metal enclosed compartment.

metal filament: The electrical conductor that glows when heated in an incandescent lamp. See Fig. M-3.

metal glaze resistors: Fixed resistors that are covered with a mixture of metal and glass.

Fig. M-3: Basic components of an incandescent lamp, including the metal filament.

metal halide lamp: A high-intensity lamp in which the light source is produced by radiation from mercury plus halides from certain other metals.

Metric system

meter: An instrument designed to measure; metric unit of linear measurement equal to 39.37 inches.

meter multiplier: A precision resistor used to extend the voltage range of a meter movement.

meter pan: A shallow metal enclosure with a round opening, through which a single kilowatt-hour meter is mounted, as the usual meter for measuring the amount of energy consumed by a particular building or other electrical system.

metric system: An international language of measurement. Its symbols are identical in all languages. The conversion factors in Fig. M-4 provide conversions from metric to English and from English to metric units, along with many other useful conversions.

Conversion Factors

Multiply	by	to obtain
acres	43,560	square feet
acres	4047	square meters
acres	1.562×10^{-3}	square miles
acres	5645.38	square varas
acres	4840	square yards
amperes	0.11	abamperes
atmospheres	76.0	cm of mercury
atmospheres	29.92	inches of merc.
atmospheres	33.90	feet of water
atmospheres	10.333	kg per sq. meter
atmospheres	14.70	pounds per sq. inch
atmospheres	1.058	tons per sq. foot
British thermal units	0.2520	kilogram-calories
British thermal units	777.5	foot-pounds
British thermal units	3.927×10^{-4}	horsepower hours

Fig. M-4: Conversion factors for metric to English, English to metric, and other units of measure.

Multiply	by	to obtain
British thermal units	1054	joules
British thermal units	107.5	kilogram-meters
British thermal units	2.928×10^{-4}	kilowatt-hours
Btu per min.	12.96	foot-pounds per sec.
Btu per min.	0.02356	horsepower
Btu per min.	0.01757	kilowatts
Btu per min.	17.57	watts
Btu per sq. ft./min.	0.1220	watts per sq. inch
bushels	1.244	cubic feet
bushels	2150	cubic inches
bushels	0.03524	cubic meters
bushels	4	pecks
bushels	64	pint (dry)
bushels	32	quarts (dry)
centimeters	0.3397	inches
centimeters	0.01	meters
centimeters	393.7	mils
centimeters	10	millimeters
centimeter-grams	980.7	centimeter-dynes
centimeter-grams	10^{-5}	meter-kilograms
centimeter-grams	7.233×10^{-5}	pound-feet
centimeters of mercury	0.01316	atmospheres
centimeters of mercury	0.4461	feet of water
centimeters of mercury	136.0	kg per sq. meter
centimeters of mercury	27.85	pounds per sq. meter
centimeters of mercury	0.1934	pounds per sq. inch
centimeters per second	1.969	feet per minute
centimeters per second	0.03281	feet per second
centimeters per second	0.036	kilometers per hour
centimeters per second	0.6	meters per minute
centimeters per second	0.02237	miles per hour

Fig. M-4: Conversion factors (cont'd).

Multiply	by	to obtain
centimeters per second	3.728×10^{-4}	miles per minute
cubic centimeters	3.531×10^{-5}	cubic feet
cubic centimeters	6.102×10^{-2}	cubic inches
cubic centimeters	10	cubic meters
cubic centimeters	1.308×10^{-6}	cubic yards
cubic centimeters	2.642×10^{-4}	gallons
cubic centimeters	10	liters
cubic centimeters	2.113×10^{-3}	pints (liq)
cubic centimeters	1.057×10^{-3}	quarts (liq)
cubic feet	62.43	pounds of water
cubic feet	2.832×10^{4}	cubic cm
cubic feet	1728	cubic inches
cubic feet	0.02832	cubic meters
cubic feet	0.03704	cubic yards
cubic feet	7.481	gallons
cubic feet	28.32	liters
cubic feet	59.84	pints (liq)
cubic feet	29.92	quarts (liq)
cubic feet per minute	472.0	cubic cm per sec.
cubic feet per minute	0.1247	gallons per sec.
cubic feet per minute	0.4720	liters per sec.
cubic feet per minute	62.4	lb of water per min.
cubic inches	16.39	cubic centimeters
cubic inches	5.787×10^{-4}	cubic feet
cubic inches	1.639×10^{-5}	cubic meters
cubic inches	2.143×10^{-5}	cubic yards
cubic inches	4.329×10^{-3}	gallons
cubic inches	1.639×10^{-2}	liters
cubic inches	0.03463	pints (liq)
cubic inches	0.01732	quarts (liq)
cubic yards	7.646×10^{5}	cubic centimeters

Fig. M-4: Conversion factors (cont'd).

Multiply	by	to obtain
cubic yards	27	cubic feet
cubic yards	46,656	cubic inches
cubic yards	0.7646	cubic meters
cubic yards	202.0	gallons
cubic yards	764.6	liters
cubic yards	1616	pints (liq)
cubic yards	807.9	quarts (liq)
cubic yards per minute	0.45	cubic feet per sec.
cubic yards per minute	3.367	gallons per second
cubic yards per minute	12.74	liters per second
degrees (angle)	60	minutes
degrees (angle)	0.01745	radians
degrees (angle)	3600	seconds
dynes	1.020×10^{-3}	grams
dynes	7.233×10^{-5}	poundals
dynes	2.248×10^{-6}	pounds
ergs	9.486×10^{-11}	kilograms
ergs	1	dyne-centimeters
ergs	7.376×10^{-8}	foot-pounds
ergs	1.020×10^{-3}	gram-centimeters
ergs	10^{-7}	joules
ergs	2.390×10^{-11}	kilogram-calories
ergs	1.020×10^{-8}	kilogram-meters
feet	12	inches
feet	0.3048	meters
feet	0.36	varas
feet	1/3	yards
feet of water	0.02950	atmospheres
feet of water	0.8826	inches of mercury
feet of water	304.8	kg per sq. meter
feet of water	62.43	pounds per sq. ft.

Fig. M-4: Conversion factors (cont'd).

Multiply	by	to obtain
feet of water	0.4335	pounds per sq. inch
foot-pounds	1.286×10^{-3}	British thermal units
foot-pounds	1.356×10^{7}	ergs
foot-pounds	5.050×10^{-7}	horsepower hours
foot-pounds	1.356	joules
foot-pounds	3.241×10^{-4}	kilogram-calories
foot-pounds	0.1383	kilogram-meters
foot-pounds	3.766×10^{-7}	kilowatt-hours
foot-pounds per minute	1.286×10^{-3}	Btu per minute
foot-pounds per minute	0.01667	foot pounds per sec.
foot-pounds per minute	3.030×10^{-5}	horsepower
foot-pounds per minute	3.241×10^{-4}	kg-calories per min.
foot-pounds per minute	2.260×10^{-5}	kilowatts
foot-pounds per sec.	7.717×10^{-2}	Btu per minute
foot-pounds per sec.	1.818×10^{-3}	horsepower
foot-pounds per sec.	1.945×10^{-2}	kg-calories per min.
foot-pounds per sec.	1.356×10^{-3}	kilowatts
gallons	8.345	pounds of water
gallons	3785	cubic centimeters
gallons	0.1337	cubic feet
gallons	231	cubic inches
gallons	3.785×10^{-3}	cubic meters
gallons	4.951×10^{-3}	cubic yards
gallons	3.785	liters
gallons	8	pints (liq)
gallons	4	quarts (liq)
gallons per minute	2.228×10^{-3}	cubic ft per sec.
gallons per minute	0.06308	liters per second
grains (troy)	1	grains (av)
grains (troy)	0.06480	grams
grains (troy)	0.04167	pennyweights (troy)

Fig. M-4: Conversion factors (cont'd).

Multiply	by	to obtain
grams	980.7	dynes
grams	15.43	grains (troy)
grams	10^{-3}	kilograms
grams	10^3	milligrams
grams	0.03527	ounces
grams	0.03215	ounces (troy)
grams	0.07093	poundals
grams	2.205×10^{-3}	pounds
horsepower	42.44	Btu per min
horsepower	33,000	foot-pounds per min.
horsepower	550	foot-pounds per sec.
horsepower	1.014	horsepower (metric)
horsepower	10.70	kg-calories per min.
horsepower	0.7457	kilowatts
horsepower	7.457	watts
horsepower (boiler)	33,520	Btu per hour
horsepower (boiler)	9,804	kilowatts
horsepower-hours	2547	British thermal units
horsepower-hours	1.98×10^6	foot-pounds
horsepower-hours	2.684×10^6	joules
horsepower-hours	641.7	kilogram-calories
horsepower-hours	2.737×10^5	kilogram-meters
horsepower-hours	0.7457	kilowatt-hours
inches	2.540	centimeters
inches	10^3	mils
inches	0.03	varas
inches of mercury	0.03342	atmospheres
inches of mercury	1.133	feet of water
inches of mercury	345.3	kg per sq meter
inches of mercury	70.73	pounds per sq ft.
inches of mercury	0.4912	pounds per sq in.

Fig. M-4: Conversion factors (cont'd).

Multiply	by	to obtain
inches of water	0.07355	inches of mercury
inches of water	25.40	kg per sq. meter
inches of water	0.5781	ounces per sq in.
inches of water	5.204	pounds per sq ft.
inches of water	0.03613	pounds per sq in.
kilograms	980,665	dynes
kilograms	10^3	grams
kilograms	70.93	poundals
kilograms	2.2046	pounds
kilograms	1.102×10^{-3}	tons (short)
kilograms-calories	3.968	British thermal units
kilograms-calories	3086	foot-pounds
kilograms-calories	1.558×10^{-3}	horsepower-hours
kilograms-calories	4183	joules
kilograms-calories	426.6	kilogram-meters
kilograms-calories	1.162×10^{-3}	kilowatt-hours
kg-calories per min.	51.43	foot pounds per sec.
kg-calories per min.	0.09351	horsepower
kg-calories per min.	0.06972	kilowatts
kilometers	10^5	centimeters
kilometers	3281	feet
kilometers	10^3	meters
kilometers	0.6214	miles
kilometers	1093.6	yards
kilowatt-hours	3415	British thermal units
kilowatt-hours	2.655×10^6	joules
kilowatt-hours	1.341	horsepower-hours
kilowatt-hours	3.6×10^6	joules
kilowatt-hours	860.5	kilogram-calories
kilowatt-hours	3.671×10^5	kilogram-meters
kilowatts	56.92	Btu per min.

Fig. M-4: Conversion factors (cont'd).

Multiply	by	to obtain
kilowatts	4.425×10^4	foot-pounds per min.
kilowatts	737.6	foot-pounds per sec.
kilowatts	1.341	horsepower
kilowatts	14.34	kg-calories per min.
kilowatts	10^3	watts
log10N	2.303	logEN or ln N
logEN or 1nN	0.4343	log10N
meters	100	centimeters
meters	3.2808	feet
meters	39.37	inches
meters	10^{-3}	kilometers
meters	10^3	millimeters
meters	1.0936	yards
miles	1.609×10^5	centimeters
miles	5280	feet
miles	1.6093	kilometers
miles	1760	yards
miles	1900.8	varas
miles per hour	44.70	centimeters per sec.
miles per hour	88	feet per minute
miles per hour	1.467	feet per second
miles per hour	1.6093	kilometers per hour
miles per hour	0.8684	knots per hour
miles per hour	0.4470	M per sec.
months	30.42	days
months	730	hours
months	43,800	minutes
months	2.628×10^6	seconds
ounces	8	drams
ounces	437.5	grains
ounces	28.35	grams

Fig. M-4: Conversion factors (cont'd).

Multiply	by	to obtain
ounces	0.0625	pounds
ounces per sq inch	0.0625	pounds per sq inch
pints (dry)	33.60	cubic inches
pints (liq)	28.87	cubic inches
pounds	444,823	dynes
pounds	7000	grains
pounds	453.6	grams
pounds	16	ounces
pounds	32.17	poundals
pounds of water	0.01602	cubic feet
pounds of water	27.68	cubic inches
pounds of water	0.1198	gallons
pounds of water per min.	2.669×10^{-4}	cubic feet per sec.
pounds per cubic foot	0.01602	grams per cubic cm.
pounds per cubic foot	16.02	kg per cubic meter
pounds per cubic foot	5.786×10^{-4}	pounds per cubic inch
pounds per cubic foot	5.456×10^{-9}	pounds per mil foot
pounds per square foot	0.01602	feet of water
pounds per square foot	4.882	kg per sq. meter
pounds per square foot	6.944×10^{-3}	pounds per sq. inch
pounds per square inch	0.06804	atmospheres
pounds per square inch	2.307	feet of water
pounds per square inch	2.036	inches of mercury
pounds per square inch	703.1	kg per sq. meter
pounds per square inch	144	pounds per sq. foot
quarts	32	fluid ounces
quarts (dry)	67.20	cubic inches
quarts (liq)	57.75	cubic inches
rods	16.5	feet
square centimeters	1.973×10^{5}	circular mils
square centimeters	1.076×10^{-3}	square feet

Fig. M-4: Conversion factors (cont'd).

Multiply	by	to obtain
square centimeters	0.1550	square inches
square centimeters	10^{-6}	square meters
square centimeters	100	square millimeters
square feet	2.296×10^{-5}	acres
square feet	929.0	square centimeters
square feet	144	square inches
square feet	0.09290	square meters
square feet	3.587×10^{-8}	square miles
square feet	0.1296	square varas
square feet	1/9	square yards
square inches	1.273×10^6	circular mils
square inches	6.452	square centimeters
square inches	6.944×10^{-3}	square feet
square inches	10^6	square mils
square inches	645.2	square millimeters
square miles	640	acres
square miles	27.88×10^6	square feet
square miles	2.590	square kilometers
square miles	3,613,040.45	square varas
square miles	3.098×10^6	square yards
square yards	2.066×10^{-4}	acres
square yards	9	square feet
square yards	0.8361	square meters
square yards	3.228×10^{-7}	square miles
square yards	1.1664	square varas
temp.(degs.C)+17.8	1.8	temp.(degs.F)
temp.(degs.F)−32	5/9	temp.(degs.C)
tons (long)	2240	pounds
tons (short)	2000	pounds
yards	0.9144	meters

Fig. M-4: Conversion factors (cont'd).

MFT: Thousands of Feet.

mho: Reciprocal of ohm.

mica: A silicate that separates into layers and has high insulation resistance, dielectric strength, and heat resistance.

MI cable: Mineral insulated, metal sheathed cable.

microfarad: A unit of capacity, being one millionth of a farad.

micrometer (mike): A tool for measuring linear dimensions accurately to 0.001 inch or to 0.01 mm.

microprocessor: The control unit of a small computer that generally has one single printed circuit to perform arithmetic and logical data.

micro structure: The structure of polished and etched metals as revealed by a microscope at a magnification of more than ten diameters.

microwave: Radio waves of frequencies above one gigahertz.

mil: A unit used in measuring the diameter of wire, equal to 0.001 inch (25.4 micrometers).

MIL: Military specification.

milli: Means 1/1000. Abbreviated m.

milliammeter: A meter that reads in the milliampere range.

milliampere: One thousandth of an ampere.

millivolt: One thousandth of a volt; .0001 volt.

mil scale: The heavy oxide layer formed during hot fabrication or heat treatment of metals.

minimum at a point: Specifications permit the thickness at one point to be less than the average.

minimum average: The specified average insulation or jacket thickness.

minimum circuit ampacity (MCA): A name plate piece of information that is used to determine the conductor size and switch rating.

mks: Meter, kilogram, second.

mm: Millimeter; 1 meter/1000.

mode: An electromagnetic field that travels in fiber.

modem: Equipment that connects data transmitting/receiving equipment to telephone lines; a word contraction of "modulator-demodulator."

modular jack: A female telephone jack that connects a telephone to the building's telephone system.

modulation: 1) The varying of a "carrier" wave by a characteristic of a second "modulating" wave. 2) To regulate by or adjust to a certain measure or proportion.

modulus of electricity: The ratio of stress (force) to strain (deformation) in a material that is elastically deformed.

mogul: A socket or receptacle (lampholder) used with large incandescent lamps of 300 watts or more.

moisture-repellent: So constructed or treated that moisture will not penetrate.

moisture-resistance: So constructed or treated that moisture will not readily injure.

molded case breaker: A circuit breaker enclosed in an insulating housing.

mole (mol): The basic SI unit for amount of substance; one mole is the amount of substance of a system that contains as many elementary entities as there are atoms in 0.012 kilograms of carbon 12.

molecule: The group of atoms that constitutes the smallest particle in which a compound or material can exist separately.

monostable: When an oscillator is in a mode of operating through just one sequence of events.

motor: An apparatus to convert from electrical to mechanical energy.

motor, capacitor: A single-phase induction motor with an auxiliary starting winding connected in series with a condenser for better starting characteristics.

motor circuit switch: A rated horsepower switch that is designed to interrupt the motor overload current.

Fig. M-5: Motor controls are offered in many variations. This starter has the option of either a push button or selector switch.

motor control: Device to start and/or stop a motor at certain temperature or pressure conditions. See Fig. M-5.

motor control center: A grouping of motor controls such as starters.

motor disconnecting means: A method by which a motor may be disconnected from the circuit such as by a switch or circuit breaker. Must be able to interrupt safely locked rotor amps.

motor-driven timer: A synchronous clock motor that is a very accurate timer.

motor effect: Movement of adjacent conductors by magnetic forces due to currents in the conductors.

motor starter: A contactor that has built-in overload protection.

mounting slots: Slots that are designed into a device to allow adjustments that can be used to make the proper alignment.

mounting strap: The metal structure built into a switch or receptacle that has slots for proper adjustment; also a specially designed strap to mount to a ceiling box to hold a light fixture.

mouse: Any weighted line used for dropping down between finished walls to attach to cable to pull the cable up; a type of vertical fishing between walls.

MPT: Male Pipe Thread.

MPX: Multiplexer.

MSDS sheets: Material Safety Data Sheets required by OSHA to be filled out on all hazardous materials.

MTW: Machine Tool Wire.

multimeter: An electrical testing instrument that measures two or more electrical quantities. See Fig. M-6.

multimode fiber: There are two types of multifibers; step-index and graded-index. How the light travels within the multifiber cable determines the use. See Fig. M-7.

multioutlet assembly: A type of surface or flush raceway designed to hold conductors and attachment plug receptacles.

Fig. M-6: An analog multimeter or VOM is a testing instrument that can test two or more quantities.

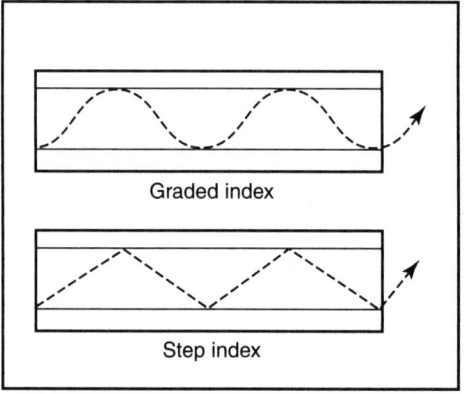

Fig. M-7: Multimode fiber (graded and step).

multiple barrier protection (nuclear power): The keeping of radioactive fission products from the public by placing multiple barriers around the reactor; for BWR these are fuel cladding, the reactor vessel, and the containment building.

multiplex: To interleave or simultaneously transmit two or more messages on a single channel.

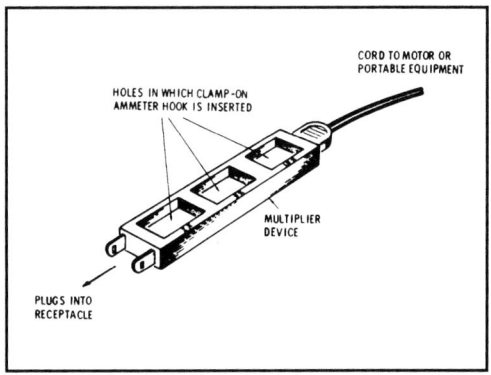

Fig. M-8: A multiplier used on an ammeter.

multiplier: A known resistance used with a multimeter or a galvanometer to increase their range. See Fig. M-8.

multiset coupler: A device that allows multiple connections to the satellite, cable, or antenna system.

multispeed motor: A motor capable of being driven at any one of two or more different speeds independent of the load. See Fig. M-9, on page 180.

multi-tap: A term used to show that a transformer has multiple taps on the secondary windings. See Fig. M-10, page on 181.

multiwire: A term generally used when more than two conductors are in a single cable. See Fig. M-11, on page 181.

mutual inductance: The condition of voltage in a second conductor because of a change in current in another adjacent conductor.

mw: Megawatt.

Mylar®: DuPont trade name for a polyester film whose generic name is oriented polyethylene terephthalate; used for insulation, binding tapes.

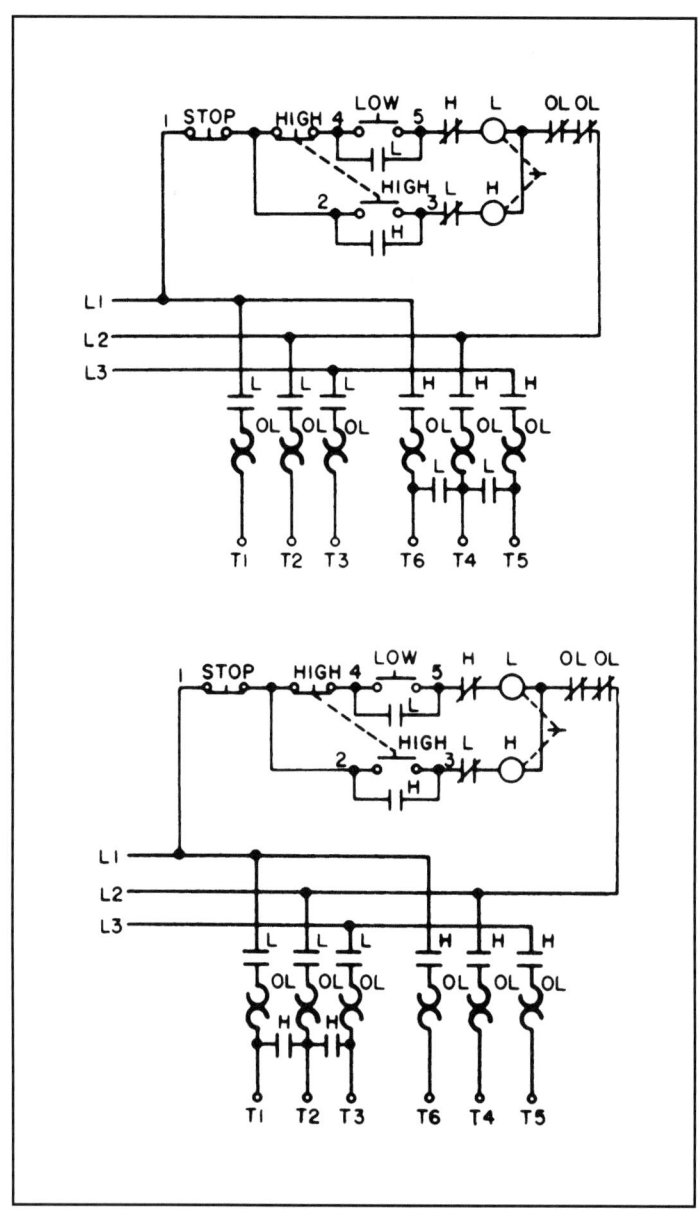

Fig. M-9: A wiring diagram showing one type of multispeed motor.

Fig. M-10: Multi-tap transformer.

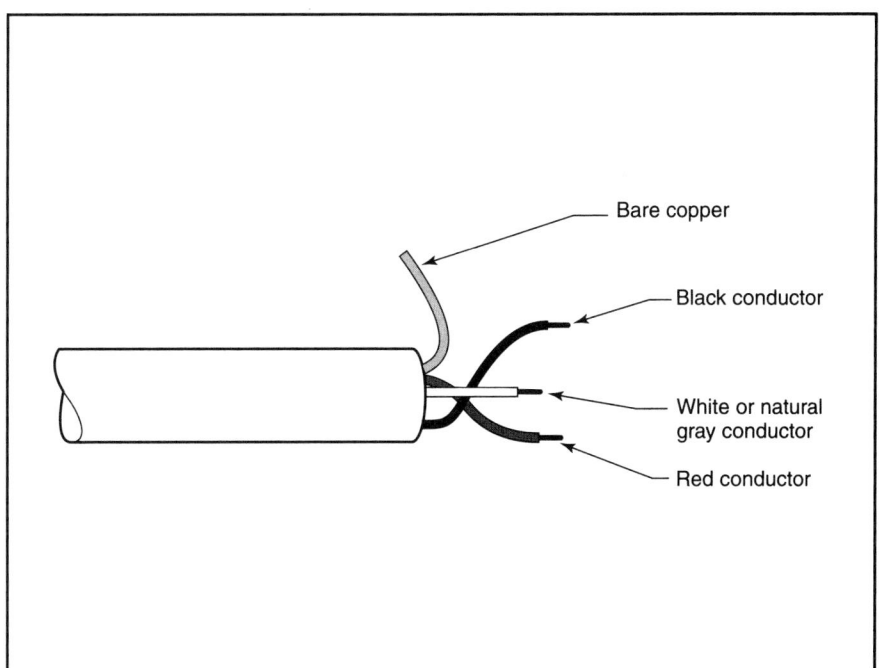

Fig. M-11: A multiwire cable has two or more conductors in a single cable.

N

N/A: 1) Not Available. 2) Not Applicable.

NAND gate: A device that will produce a low output when both inputs are high. See Fig. N-1.

nano: Meaning 1/1,000,000,000; abbreviated n.

nanometer: A meter to measure the wavelength of light.

National Electrical Code® (NEC®): A set of rules governing the selection of materials, quality of workmanship, and precautions for safety in the installation of electrical wiring. The *NEC®*, originally prepared in 1897, is frequently

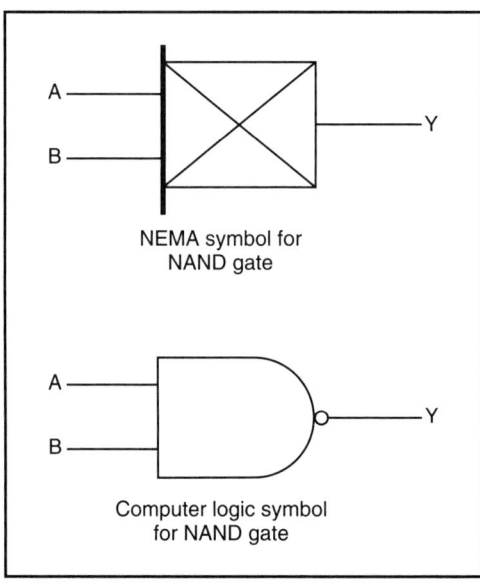

Fig. N-1: NAND gate symbols for NEMA and computer logic.

revised to meet changing conditions, such as improved equipment and materials and new fire hazards. The code is the result of the best efforts of electrical engineers, manufacturers of electrical equipment, insurance underwriters, firefighters, and other concerned experts through the country.

The *NEC*® is published by the NFPA (National Fire Protection Association), Batterymarch Park, Quincy, MA. It contains specific rules and regulations intended to help in "the practical safeguarding of persons and property from hazards arising from the use of electricity." The *NEC*® contains provisions considered necessary for safety. Compliance therewith and proper maintenance will result in an installation essentially free from hazard, but not necessarily efficient, convenient, or adequate for good service for future expansion of electrical use.

natural convection: Movement of a fluid or air caused by temperature change.

NBR: Nitrite-Butadiene Rubber: synthetic rubber.

NBS: National Bureau of Standards.

NC: Normally Closed.

needle nose pliers: Slim pliers with a long nose that are excellent for making loops in conductors for easy connection to electrical devices. Most have a blade edge to cut small conductors.

negative (–): 1) Connected to the negative terminal of a power supply. 2) One of the polarities of electricity.

negative charge (–): The electrical polarity sign that indicates an abundance of negative electrons.

negative conductor: A conductor leading from the negative terminal.

negative-ion: An atom that has negative ion charge.

negative logic: 0 represents the voltage that generally is represented by 1 in binary logic.

negative plate: 1) In a storage cell, the spongy lead plate which, during discharge, is the negative plate or terminal. 2) In a primary cell, the carbon, copper, platinum, etc. is the negative electrode.

negative resistance: Current will decrease with the increase of voltage.

negative side of circuit: The conducting path of a circuit from the current-consuming device back to the source of supply.

NEMA: National Electrical Manufacturers Association.

NEMA size: A standard that NEMA has established for horsepower, current voltage, and interrupting characteristics for electrical controller devices.

neon test light: A small neon light with two test leads. When the leads are touched to a hot and a neutral, the neon light will glow if power is present.

neoprene: An oil-resistant synthetic rubber used for jackets; originally a DuPont trade name, now a generic term for polychloroprene.

network: An aggregation of interconnected conductors consisting of feeders, mains, and services.

network limiter: A current limiting fuse for protecting a single conductor.

neutral: The element of a circuit from which other voltages are referenced with respect to magnitude and time displacement in steady state conditions.

neutral block: The neutral terminal block in a panelboard, meter enclosure, gutter, or other enclosure in which circuit conductors are terminated or subdivided. See Fig. N-2.

neutral bus bar: A neutral bar that is in electrical distribution panels or in busways. It must be large enough to carry the computed maximum neutral current.

neutral plane: When a motor's armature windings are parallel to the magnetic lines of flux. At this point zero voltage is being produced.

neutral sizing: The calculation of the appropriate size of the neutral conductor. This is generally designated by the *NEC®*.

Fig. N-2: The interior of a panelboard showing a neutral block.

neutral wire: A grounded conductor in an electrical system that does not carry current until the system is unbalanced. Neutral conductors must have sufficient capacity for the current that they might have to carry under certain conditions. In a single-phase, three-wire system, the middle leg or neutral wire carries no current when the loads on each side between neutral are equal or balanced. If, however, the loads on the outside wires become unequal, the difference in current flows over the neutral wire.

When a 120/240-V single-phase service is used, it is highly desirable for the 120-V loads to be balanced across both sides of the service. The neutral wire can then be smaller than the two hot or ungrounded wires. The *NEC*® permits the reduction of the neutral to the size that will carry the maximum unbalanced load between the neutral and any one ungrounded conductor.

A general rule of thumb is to reduce the neutral by not more than two standard wire sizes. A further demand factor of 70 percent may be applied in reducing the neutral wire for that portion of the unbalanced load that is in excess of 200 A. However, if 50 percent or more of the load consists of electric-discharge lamp ballasts, the neutral will be the same size as the ungrounded conductors.

In a three-phase, four-wire system, a three-wire branch circuit consisting of two phase wires and one neutral wire has a neutral or grounded wire that carries approximately the same current as the phase conductors. Therefore, it should be the same size as the phase conductors. However, three-phase, four-wire systems generally supply a mixed load of lamps, motors, and other appliances. Motors and similar three-phase loads connected only to the phase wires cannot throw any load onto the neutral, and such three-phase loads can be disregarded in calculating the necessary capacity for the neutral conductor.

neutron: Subatomic particle contained in the nucleus of an atom: electrical neutral. The neutron is located in the central mass, or nucleus, of the atom and possesses no electrical charge. Due to this, the neutron is not deflected by magnetic or electric fields, and its interaction with matter is mainly by collision. See Fig. N-3, on page 186.

newton: A unit of measurement that expresses the force of attraction or repulsion between two charged bodies. This force varies directly with the product of the individual charges, and at the same time, varies inversely with the square of the distance between charges.

NFPA (National Fire Protection Association): An organization to promote the science and improve the methods of fire protection that sponsors various codes, including the *National Electrical Code*®.

niche fixture: A below-water type of light fixture generally used around swimming pools. See Fig. N-4.

nickel plating: The depositing of a coating of nickel on a metallic surface. Accomplished by immersion in a nickel salt bath through which an electric current of low voltage is passed.

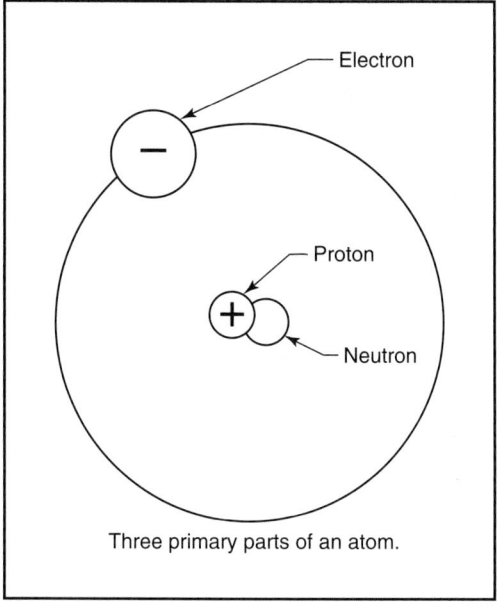

Fig. N-3: A neutron is one of the three primary parts of an atom.

nineteen hundred box: A commonly used term to refer to any 2-gang 4-inch square outlet box used for two wiring devices or for

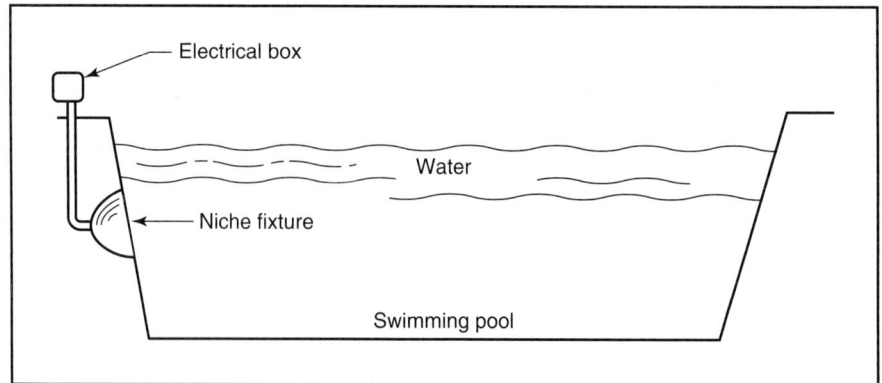

Fig. N-4: A niche fixture is a swimming pool luminaire (lighting fixture).

one wiring device with a single-gang cover where the number of wires requires this box capacity. See Fig. N-5.

nipple: A threaded pipe or conduit of less than two feet length.

NMC cable: Non-Metallic sheathed Cable that is factory assembled. The outer sheath is moisture resistant and flame retardant.

NO: Normally open.

node: A junction of two or more branches of a network.

Fig. N-5: A four-inch square or nineteen hundred box.

node current equations: Kirchhoff's current law gives a mathematical expression at a junction or node.

node voltage: A part of Kirchhoff's law. The voltage at a node with respect to some reference point in the circuit.

no-load current: The current that flows through a device or circuit when the same device or circuit is delivering zero-output current. For example, when the primary of a transformer is connected to an ac source, a voltage appears at the secondary winding even when it is not connected to a load. While there is no current flow from the output in this condition, a small amount of current flows in the primary winding to sustain transformer operation. Here, the no-load current is the measured drain from the primary input voltage source. See Fig. N-6, on page188.

nominal: Relating to a designated size that may vary from the actual.

nominal rating: The maximum constant load that may be increased for a specified amount for two hours without exceeding temperature limits specified from the previous steady state temperature conditions; usually 25 or 50 percent increase is used.

nomograph: A chart or diagram with which equations can be solved graphically by placing a straightedge on two known values and reading where the straightedge crosses the scale of the unknown value.

Fig. N-6: Even with the load disconnected from the transformer, there is still some current drain in the primary winding.

nonautomatic: Used to describe an action requiring personal intervention for its control.

noncode installation: A system installed where there are no local, state, or national codes in force.

non-coincident load: Electrical loads that are not used at the same time.

non-combustible: Nonflammable, not easy to burn.

nonconductor: Any substance that does not allow electricity to pass through it.

no-niche fixture: A fixture mounted on a bracket, installed not less than 18 inches below normal water level, which must have a cord from the fixture to a deck box at the top of the pool.

noninductive circuit: A circuit in which the magnetic effect of the current flowing has been reduced by one of several methods to a minimum or to zero.

noninductive resistance: Resistance free from self-induction.

noninductive winding: A winding so arranged that the magnetic field set up by the current flowing in one half of the coil is neutralized by the magnetic field set up by the current flowing in the opposite directions in the second half.

nonmetallic-sheathed (NM) cable: A type of cable that is popular for use in residential and small commercial wiring systems. In general, it may be used for both exposed and concealed work in normally dry locations.

Type NM cable must not be installed where exposed to corrosive fumes or vapors, nor embedded in masonry, concrete, fill, or plaster, nor run in shallow chases in masonry or concrete and finished or covered with plaster or similar finish. This cable must not be used as a service-entrance cable, in commercial garages, theaters and assembly halls, motion picture studios, storage battery rooms, hoistways, hazardous locations, or embedded in poured cement, concrete, or aggregate.

For use in wood structures, holes are bored through wood studs and joists, and the cable is then pulled through these holes to the various outlets. The holes normally give sufficient support, providing they are not over 4 feet on center. When no stud or joist support is available, staples or some similar supports are required for the cable. The supports must not exceed 4.5 feet and must be within 12 inches of each outlet box or other termination point. See *Romex*.

non-power limited circuits: A 600 volt fire alarm system that has limited voltage but is not limited to the amount of current.

non-prewired: Requires additional wiring.

nonreversing: Only in one direction.

non-time-delay fuse: A fuse that has no time delay element as part of its construction and reacts when current exceeds preset levels.

NOR gate: Provides low output when either or both inputs are high. See Fig. N-7, on page 190.

normal charge: The thermal element charge that is part liquid and part gas under all operating conditions.

normally closed: The relay contacts are closed when the coil is de-energized.

normally open: The relay contacts are open when the coil is de-energized.

notation: A method of placing information on blueprints that can include symbols, quantities, numbers, or other facts or values.

not inherently-limited transformer: A type of chime transformer. Rated at 100 amps or less.

NOT logic: When the control signal is off, there is an output. The output remains on only if the control signal remains off.

NPT: National tapered Pipe Thread.

NR: 1) Nonreturnable reel; a reel designed for one-time use only. 2) Natural rubber.

NRC: (Nuclear Regulatory Commission): The Federal agency for atomic element usage; formerly AEC.

NRTL: Nationally Recognized Testing Laboratory.

NSD (Neutral Supported Drop): A type of service cable.

n-type material: A crystal that has been doped with material from another element that has more electrons in its outer shell than the crystal has.

nucleus: An atom's center, which contains neutrons and protons.

nuisance static charge: A static charge that causes interference in a circuit.

nut driver: A hand tool with a screwdriver type handle that is used to tighten small nuts and bolts. Also called spin tight nut driver.

Nylon®: This is the DuPont trade name for polyhexamethyleneadipamide which is the thermoplastic used as insulation and jacketing material.

Fig. N-7: NOR gate symbols for NEMA and computer logic.

O

oblique drawing: It has three axes and two of the axes are always at right angles to each other and the third is generally at a 30° or 45° to the horizontal. (Similar to an isometric drawing.)

OC: Overcurrent.

OD: Outside Diameter.

O.D. size: The outside (diameter) measurement of a conduit.

OF: Oxygen Free.

off-delay timers: A timing device that responds immediately when energized but has a slow return to normal position when de-energized.

offgassing: Percentage of a gas released during combustion.

offset (pipe bending): Two bends that have the same degree of offset and are in the same plane but in opposite directions.

offset bar hanger: A hanger that connects to the bottom of a ceiling joist that is designed for plaster type ceilings.

ohm: The derived SI unit for electrical resistance or impedance; one ohm equals one volt per ampere.

ohmic resistance: The dc resistance of an inductor.

Fig. O-1: Ohmmeter used to check the resistance to a small control transformer.

ohmmeter: An instrument for measuring resistance in ohms. An ohmmeter being used to trouble-shoot a small control transformer is shown in Fig. O-1. The ohmmeter's pointer deflection is controlled by the amount of battery current passing through the moving coil. Before measuring the resistance of an unknown resistor or electrical circuit, the ohmmeter must first be calibrated. If the value of resistance to be measured can be estimated within reasonable limits, a range is selected that will give approximately half-scale deflection when the resistance is inserted between the probes. If the resistance is unknown, the selector switch is set on the highest scale. Whatever range is selected, the meter must be calibrated to read zero before the unknown resistance is measured.

ohm resistance: A dc circuit is said to have a resistance of one ohm when one volt will produce a current of one ampere through it.

Ohm's law: Mathematical relationship between voltage, current, and resistance in an electric circuit. Ohm's law states that the current flowing in a circuit is proportional to the voltage and inversely proportional to the resistance or opposition. See Fig. O-2.

OI (Official Interpretation): A now obsolete term used for an interpretation of the National Electrical Code made to help resolve a specific problem between an inspector and an installer. *Formal Interpretations* is the term now used to promote uniformity of interpretation and application of the provisions in the National Electrical Code. The procedures may be found in the "NFPA Regulations Governing Committee Projects."

oil-: The prefix designating the operation of a device submerged in oil to cool or quench or insulate.

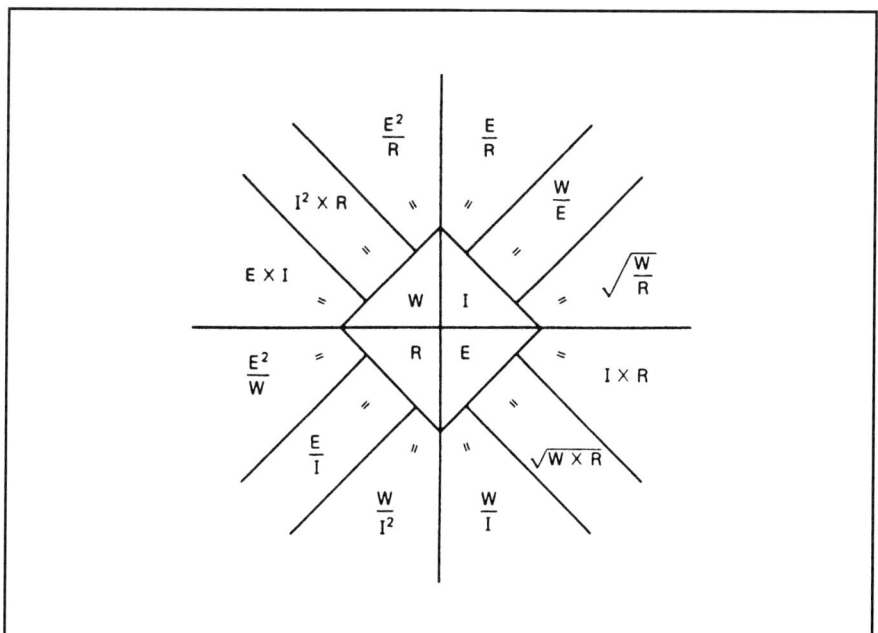

Fig. O-2: Summary of Ohm's law.

oil-proof: A device or apparatus designed so that the accumulation of oil or vapors will not prevent safe, successful operation.

oil-tight: Construction preventing the entrance of oil or vapors not under pressure.

OL *(nuclear):* Operating License.

OL: Overload.

on-delay timers: The contacts of a relay that respond slowly when the coil is energized but return back to their normal position when de-energized.

one-line diagrams: The use of single lines and symbols to illustrate an electrical circuit.

opacity: The rate at which light rays can pass through an object.

OP-amp: Operational amplifier. Has a high impedance and needs very little current to operate.

open circuit: A circuit that is energized but not allowing useful current to flow. For example, a switch installed in an energized circuit can be opened so as to stop the flow of current through the circuit.

open delta system: A system that uses only two transformers instead of three to furnish three-phase power. See Fig. O-3.

Fig. O-3: Open delta system used by only two transformers.

opening: An old electrical term meaning a receptacle and/or light fixture outlet.

opening time: The period between which an activation signal is initiated and switch contacts part.

open wiring: Exposed electrical conductors.

operator: An electrical device that is operated by an individual who is controlling the circuit by pressing, pulling, or rotating it.

opposite side: The side opposite the right angle in a triangle.

optical amplifier: Increases light intensity without converting it to an electrical signal.

optical fiber cable: A cable that has a light-carrying core that is surrounded by cladding, which keeps the light from escaping from the core. See Fig. O-4.

optical power: Measured in radiant energy per unit of time. Example: 0 dB = 1 Watt.

optimization: The procedure used in the design of a system to maximize or minimize some performance index.

opto-isolater: One side of the load side of the solid-state relay is optically isolated from the control side of the relay.

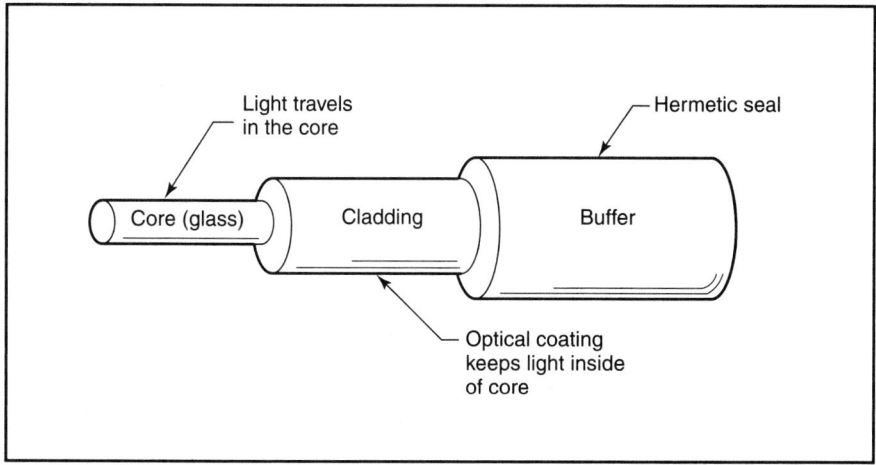

Fig. O-4: Optical fiber cable is a light-carrying cable.

orbit: The path around the nucleus of an atom taken by the electrons. See *atom*, *shell*, and *valence*.

organic: Matter originating in plant or animal life, or composed of chemicals of that origin.

OR gate: A device that has a high output when either or both of the inputs are high. See Fig. O-5.

orifice: Accurate size opening for controlling fluid flow.

OR logic: An output of logical 1 when one or more inputs are 1.

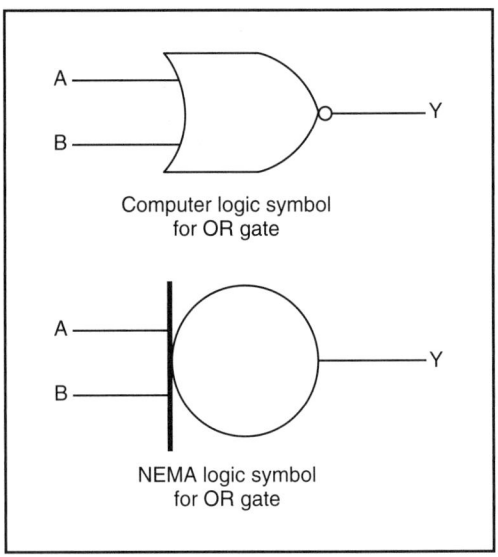

Fig. O-5: OR gate symbols for NEMA and computer logic.

oscillation: The variation, usually with time, of the magnitude of a quantity that is alternately greater and smaller than a reference.

oscillator: A device that produces an alternating or pulsating current or voltage electronically.

oscillograph: An instrument primarily for producing a graph of rapidly varying electrical quantities.

oscilloscope: An instrument primarily for making visible rapidly varying electrical quantities; oscilloscopes function similarly to TV sets. The principle components of a basic oscilloscope include a cathode-ray tube (CRT), a sweep generator, deflection amplifiers (horizontal and vertical), a power supply, and suitable controls, switches, and input connectors for proper operation. Fig. O-6 shows an oscilloscope.

OSHA (Occupational Safety and Health Act): Federal Law #91-596 of 1970 charging all employers engaged in business affecting interstate commerce to be responsible for providing a safe working place; it is administered by the Department of Labor; the OSHA regulations are published in Title 29, Chapter XVII, Part 1910 of the CFR and the Federal Register.

Fig. O-6: Oscilloscopes are used to show rapidly varying electrical quantities.

osmosis: The diffusion of fluids through membranes.

ought sizes: An expression referring to conductors of sizes No. 1/0, 2/0, 3/0, or 4/0. Sometimes called "naught."

outage: The condition resulting when a component is not available to perform its intended function.

outdoor: Designed for use out-of-doors.

outgassing: Dissipation of gas from a material.

outlet: A point on the wiring system at which current is taken to supply utilization equipment.

outlet box: A box installed in an electrical wiring system, from which current is taken to supply some apparatus or device.

outlet tester: A receptacle tester that has different colored LEDs and the way that the lights light up tells the electrician if the outlet is properly installed.

outline lighting: An arrangement of incandescent lamps or gaseous tubes to outline and call attention to certain features, such as the shape of a building or the decoration of a window.

out-of-phase voltage: Where two-phase voltages are out-of-phase with each other in a wye connection. Example: If each phase winding is 120 volts, the two-phase winding connected together is 208 volts not 240 volts.

output: 1) The energy delivered by a circuit or device. 2) The terminals for such delivery.

output pulse: A voltage or current produced at the output of a device that lasts a very short period of time.

output states indicator: Output indicators are generally analog meters, digital meters, stepping motor signals, variable voltage outputs, and variable current outputs.

oven: An enclosure and associated sensors and heaters for maintaining components at a controlled and usually constant temperature.

overcurrent: Any current that is more than the rated current of conductors and/or equipment.

overcurrent protection: De-energizing a circuit whenever the current exceeds a predetermined value; the usual devices are fuses, circuit breakers, or magnetic relays.

overhead cost: The cost of operating a business, as for rent, interest on investment, maintenance and depreciation of equipment, etc.; over and above the actual cost of labor and material.

overhead service: An electrical service that runs from a utility pole to a building in overhead cables.

overload: Load greater than the load for which the system or mechanism was intended. A fault, such as a short circuit or ground fault, is not an overload.

overload protection: A device that detects excess current over time to electrical equipment and opens the circuit.

overload relay reset: Push-button type of device that resets an overload relay after it has tripped.

overload switch: An automatic switch for breaking a circuit in case of an overload.

over-time: Time spent in working beyond specified working hours.

overvoltage (cable): Voltages above normal operating voltages, usually due to: a) switching loads on/off; b) lighting; c) single phasing.

oxidize: 1) To combine with oxygen. 2) To remove one or more electrons. 3) To dehydrogenate.

oxygen bomb test: Aging a rubber sample in a pressure container in a pure oxygen environment not deteriorated by ozone.

oxygen index: A test to rate flammability of materials in a mixture of oxygen and nitrogen.

ozone: An active molecule of oxygen that may attack insulation; produced by corona in air.

ozone resistance tests: A test for rubber, not used now because the synthetic rubbers do not deteriorate from ozone.

P

P102: A cable legend signifying acceptance for listing by Pennsylvania Department of Mines (POM).

pad: A coil of tape.

paddle fan: A ceiling fan that has several large paddles and is used to circulate air in a room or designated area.

pad-mounted: A shortened expression for "pad-mount transformer," which is a completely enclosed transformer mounted outdoors on a concrete pad, without need for a surrounding chain-link fence around the metal, box-like transformer enclosure. See Fig. P-1.

pan: A sheet metal enclosure for a watt-hour meter, commonly called a "meter pan."

panel: A unit for one or more sections of flat material suitable for mounting electrical devices.

Fig. P-1: Typical pad-mounted transformer.

Parallel Circuit

panelboard: A single panel or group of panel units designed for assembly in the form of a single panel; includes buses and may come with or without switches and/or automatic overcurrent protective devices for the control of light, heat, or power circuits of individual as well as aggregate capacity. It is designed to be placed in a cabinet or cutout box that is in or against a wall or partition and is accessible only from the front. See Fig. P-2.

Fig. P-2: Typical electrical panelboard.

paper-lead cable: One having oil impregnated paper insulation and a lead sheath.

parallax: This occurs when the scale of an electrical instrument is not in correct alignment.

parallel: Lying side by side; extending in the same direction and equidistant at all points.

parallel circuit: A circuit having a common return between which two or more pieces of apparatus are connected, each receiving a separate portion of the current flow from the common feed. See Fig. P-3.

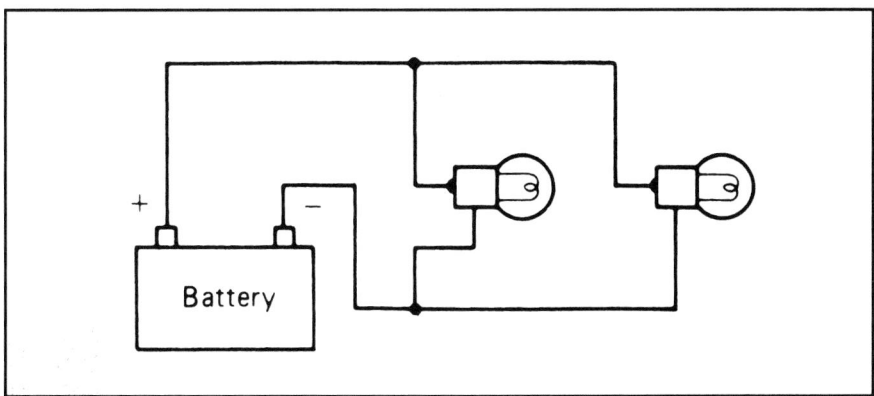

Fig. P-3: A circuit with lamps connected in parallel.

parallel-connected transformer: Two or more transformers having their primary windings connected to the same source of supply in such a manner that the impressed voltage in each case is the same as that of the line.

parallelogram: A four-sided figure whose opposite sides are equal in length and parallel to each other.

parameter: A variable given a constant value for a specific process or purpose.

par lamp: Parabolic aluminized reflector lamp used when a sealed beam is required for stage, studio, or outdoor lighting.

part-wind starting: A type of motor winding that has two or more circuits in the stator winding. The individual circuits are in series with each other when starting the motor and in parallel when the motor is running.

pascal: The derived SI unit for pressure or stress: one pascal equals one newton per square meter.

patch: To connect circuits together temporarily.

payoff: The equipment to guide the feeding of wire.

PB: Pushbutton; pull box.

PC boards: A non-conductive type of material that has conductive paths fixed to it.

PE: 1) Polyethylene. 2) Professional Engineer.

peak inverse voltage: The maximum reverse bias voltage that a diode can withstand.

peak load: The heaviest load that a generator or system is called on to supply at regular intervals, as once every 24 hours.

peak value: The largest instantaneous value of a variable.

penciling: The tapering of insulation to relieve electrical stress at a splice or termination.

penetration (nuclear): A fitting to seal pipes or cables through the containment vessel wall.

pentode: An electron tube with five electrodes or elements.

period: The minimum interval during which the same characteristics of a periodic phenomenon recur.

permalloy: An alloy of nickel and iron that is easily magnetized and demagnetized.

permanent magnet: A piece of magnet steel that retains the acquired property of attracting other pieces of magnetic material after being under the influence of a magnetic field.

permanent magnet moving-coil meter: The basic movement used in most measuring instruments for servicing electrical equipment. The basic movement consists of a stationary permanent magnet and a movable coil. When current flows through the coil, the resulting magnetic field reacts with the magnetic field of the permanent magnet and causes the coil to rotate. The greater the intensity of the current flow through the coil, the stronger the magnetic field produced; and the stronger the magnetic field, the greater the rotation of the coil.

permanent-split capacitor motor: A single-phase induction motor that uses the same capacitance for both the starting and the running of the motor. See Fig. P-4.

permeability: The ability of a core of magnetic material to conduct lines of force. Some core materials have higher permeabilities than others and thus, used with a given winding, they provide

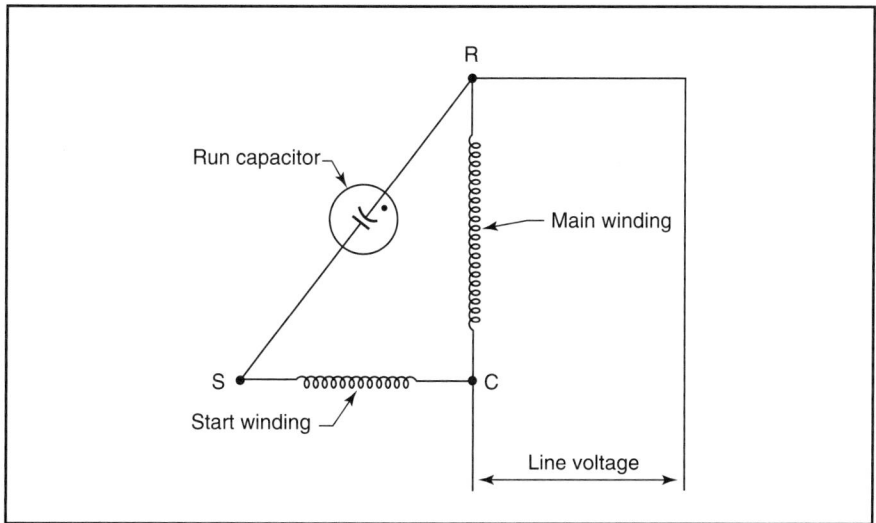

Fig. P-4: A schematic of the windings in a permanent-split capacitor motor.

inductors having greater inductance values. More precisely, the permeability of a certain magnetic material is the ratio of the flux produced with that magnetic material as the core to the flux produced with air as the core.

It is important to be aware of the fact that the permeability of a given magnetic material varies with the magnetizing force in ampere-turns per unit of core length that is applied to the core. The permeability also depends on the amount and direction of any magnetic flux that might already exist.

permissible mine equipment: Equipment that has been formally accepted by the Federal agency.

per-unit quantity: The ratio of the actual value to an arbitrary base value of a quantity.

PES: Power Engineering Society of IEEE.

PF: Power Factor.

pH: An expression of the degree of acidity or alkalinity of a substance; on the scale of 1 to 10, acid is under 7, neutral is 7, alkaline is over 7.

phase: The fractional part "t/p" of the period through which a quantity has advanced, relative to an arbitrary origin. The time instant when the maximum, zero, or other relative value is attained by an electric wave.

phase angle: The measure of the progression of a periodic wave in time or space from a chosen instant or position.

phase conductor: The conductors other than the neutral.

phase converter: A device that will permit the operation of a three-phase induction motor from a single-phase power source. Since most commercial suppliers of power place limits on the size of single-phase motors they can serve, such limitations may also apply to three-phase motors supplied through a phase converter. The power supplier should be consulted regarding the size of converter that can be operated in a particular location.

Application of phase converters to easy-starting loads usually involves no particular problems. Such loads would include large fans and centrifugal and turbine-type irrigation pumps. For equipment that requires high starting torque or is subject to wide load fluctuation, application of a phase converter should be made only after consulting the manufacturer of the three-phase motor. Such equipment would include compressors, pumps, and barn cleaners that start under load and feed grinders and blowers on which load may vary due to uneven feeding.

Phase converters may be divided into two general types—the static converter and the rotating transformer converter. Static converters are subdivided into several types, among which are autotransformer converters, series-winding converters, and multimotor converters. Some of these must be matched in horsepower rating to the motor to be driven; that is, a 5-hp converter for a 5-hp motor, etc. The multimotor converter, as its name implies, will operate two or more motors.

The rotating transformer-type converter should have a horsepower rating as large as that of the largest motor to be driven. Additional smaller motors may be supplied by the same converter.

The starting current of a converter (three-phase motor combination) is likely to be less than for a comparable single-phase motor. By the same token, the starting torque of the motor-converter combination is likely to be less than for a similar-sized single-phase motor or for a three-phase motor operated from a three-phase supply. Likewise, the motor connected through a converter may have very little short-time overload capacity.

Overcurrent protection may be difficult to provide because of the longer starting period and the unbalanced currents that occur in the motor windings under overload conditions. Power factor is likely to be near 100 percent at rated load and slightly leading when idling on the line.

phase current: The current through each load in wye or delta circuits.

phase difference: The difference in the timing of current and voltage outputs.

phase-failure protection: A solid state device that monitors and shuts down a motor when there is a loss of a phase leg or a phase reversal.

phase leg: One of the phase conductors (an ungrounded or "hot" conductor) of a polyphase electrical system.

phase loss: When three-phase drops one of the phases, causing the motor to run on single-phase.

phase meter: A meter that indicates the frequency of the circuit to which it is attached; a frequency meter.

phase out: A procedure by which the individual phases of a polyphase circuit or system are identified; such as to "phase out" a 3-phase circuit for a motor in order to identify phase A, phase B, and phase C to know how to connect them to the motor to get the correct phase rotation so the motor will rotate in the desired direction.

phase protection: A means of preventing damage in an electric motor through overheating in the event a fuse blows or a wire breaks when the motor is running. This is necessary because a motor will continue to operate on single phase even if some part of the device is damaged. In order to provide the protection, phase-failure and -reversal relays are used. Phase-failure relays are available in a number of designs, some of which are quite complex devices.

One type of relay utilizes coils that are connected in two lines of the three-phase supply. The currents in the coils cause a rotating magnetic field to be set up to turn a copper disc clockwise. This movement, which is known as *torque*, results from one polyphase torque turning the disc clockwise and one single-phase torque turning the disc counterclockwise. The disc is kept from turning by a projection resting against a stop.

In the event that the disc rotates counterclockwise, the projecting arm causes a toggle mechanism to open line contacts, thus removing the motor from the line and preventing any damage. In the event that one line opens, the polyphase torque disappears, the single-phase torque rotates the disc counterclockwise, and the motor is again removed from the line, thus preventing overheating.

Some complex types of phase-failure relays are commonly used in situations where it is necessary to protect not only a device, but also any persons or machines from the dangers involved in open-phase or reversed-phase sequence conditions. In some cases, a phase-failure relay may consist of a static, current-sensitive network that is connected in series with the line and a switching relay connected in the coil circuit of the starter. This sensing network serves to monitor the line current in the motor. In the event one phase opens, the sensing network detects it, causing the relay to open the starter coil current, which disconnects the motor from the line. This type of phase-failure relay usually contains a built-in delay of five cycles, which prevents nuisance drops that are caused by transient line fluctuations.

phase reversal: An improper phase sequence, which can lead to damage to electric motors.

phase-reversal protection: See *phase-failure protection.*

phase rotation relay: A relay that operates with the direction of the phase rotation.

phase sequence: The order in which the successive members of a periodic wave set reach their positive maximum values; a) zero phase sequence—no phase shift; b) plus/minus phase sequence—normal phase shift.

phase shift: The absolute magnitude of the difference between two phase angles.

phase-to-phase voltage: In a delta-connection, the line voltage and the phase voltage are the same. In a wye-connection, the line voltage is 1.732 times greater than the phase voltage.

phase voltage: The voltage reading across a single-phase winding of a motor or transformer. This reading is phase to phase in contrast to phase-to-neutral or phase-to-ground.

phasing: The interconnection of the windings of a motor, transformer, or generator so that the phase relationships are correct.

phasor: A line representing ac current or voltage at a specific time.

phasor quantity: A complex algebraic expression for sinusoidal wave.

photocell: A device in which the current-voltage characteristic is a function of incident radiation (light).

photoconductive: Changing the resistance or conductance by increasing or decreasing the light energy level.

photodiode: A diode that can be switched on or off by a light.

photoelectric control: A control sensitive to incident light. See Fig. P-5 on page 208.

photoelectricity: A physical action wherein an electrical flow is generated by light waves.

Fig. P-5: Data for a photoelectric switch.

photoelectric-type smoke detector: See *photoelectronic smoke detector.*

photoelectronic smoke detector: Designed with a light sensor that responds when smoke reduces the amount of light.

photometer: An instrument for measuring the intensity of light or for comparing the relative intensity of different lights.

photon: An elementary quantity of radiant energy (quantum).

photovoltaic: Voltage being produced from a light source.

pi (π): The ratio of the circumference of a circle to its diameter.

pick: The grouping or band of parallel threads in a braid.

pickle: A solution or process to loosen or remove corrosion products from a metal.

pickup value: The minimum input that will cause a device to complete a designated action.

pico: Meaning one million. Abbreviated p.

picocoulomb: 10^{-12} coulombs.

Piezoelectric effect: Some materials become electrically polarized when they are mechanically strained; the direction and magnitude of the polarization depends upon the nature, amount and the direction of the strain; in such materials the reverse is also true in that a strain results from the application of an electric field.

piezoresistor: A pressure sensor that can detect either a vacuum, differential pressure, or absolute pressure.

pigtail: A flexible conductor attached to an apparatus for connection to a circuit.

PILC cable: Paper Insulated, Lead Covered.

pilot device: Used to control larger controlling devices. Generally uses a small amount of current.

pilot lamp: A lamp that indicates the condition of an associated circuit.

pilot wire: An auxiliary insulated conductor in a power cable used for control or data.

pin gauge: An insulation thickness gauge having a pin that will fit the groove left by a wire strand.

pipe dies: Tools that cut the male thread on a piece of pipe or metal rod. They come in many different sizes for both the pipes and the rods.

pipe taps: Tools that cut the female thread in a fitting used for piping or for a metal rod coupling.

pipe vise: Type of vise for securing conduit during cutting, reaming, and threading.

pitch diameter: The diameter through the center of a layer in a concentric layup of a cable or strand.

pitting: Small cavities in a metal surface.

plane vector: Graphically shows two or more vectors at the same time and their relationship.

plasma: A gas made up of charged particles.

plaster and lath: An old technique whereby wooden laths were plastered over to make the ceiling or walls of the building.

plaster ears: Adjustable ears mounted on electrical boxes that allow for adjustment for a flush finish.

plastic: Pliable and capable of being shaped by pressure.

plastic deformation: Permanent change in dimensions of an object under load.

plasticizer: Agent added to plastic to improve flow and processability and to reduce brittleness.

plate: A rectangular product having thickness of 0.25 inch or more.

plate holes: Holes that are drilled through the top plates to allow access to the hollow core of a wall for installation of electric cable.

plating: Forming an adherent layer of metal on an object.

plenum: Chamber or space forming a part of an air conditioning system.

plenum cable: Cable that is specifically rated to be installed in air handling plenums.

plowing: Burying cable in a split in the earth made by a blade.

plug: A male connector for insertion into an outlet or jack.

plug fuse: A type of fuse that is held in position by a screw-thread contact instead of spring clips, as is the case with a cartridge fuse.

plugging: Braking an induction motor by reversing the phase sequence of the power to the motor. This reversal causes the motor to develop a counter torque, which results in the exertion of a retarding force. Plugging is used to secure both rapid stop and quick reversal.

Because it is possible for motor connections to be reversed when the motor is running, control circuits should be designed specifically to prevent this from occurring when it is undesirable. However, there are a number of factors that must be considered and investigated thoroughly when it is desired to have this type of operation. It may be necessary to have methods of limiting maximum permissible currents, particularly in situations with repeated operations and also with dc motors. The machine under consideration should be carefully investigated in order to ensure that this type of action will not do damage over an extended period of time.

plug-in relay: A relay that plugs into a connecting base.

plumb-and-chalkline: A tool that marks straight lines perpendicular to the surface that is being laid out.

plumb bob: A cone-like weight attached to a plumb line to ensure exact layout points between ceiling and floor.

pneumatic timer: A widely versatile timer that uses an orifice and bellows. The air flow into the bellows is controlled by a needle valve.

PN-junction: The area between the P-type and the N-type material on a semiconductor.

point-to-point: A data communications system that has connections between two points as to multipoints.

polarity: 1) Distinguishing one conductor or terminal from another. 2) Identifying how devices are to be connected, such as positive (+) or negative (−).

polarization: Gas ions are created around the positive terminal of a battery and when enough gas ions gather, the cell will become useless.

Polarization Index: Ratio of insulation resistance measured after 10 minutes to the measure at 1 minute with voltage continuously applied.

polarized plugs: Electrical plugs that have one prong larger than the other. See Fig. P-6.

polarized receptacle: Outlets that have one slot larger than the other and receive a polarized plug.

pole: 1) That portion of a device associated

Fig. P-6: A polarized plug is one that has one prong larger than the other.

exclusively with one electrically separated conducting path of the main circuit of device. 2) A supporting circular column.

poly: Polyethylene.

polychloroprene: Generic name for neoprene.

polycrystalline: Pertaining to a solid having many crystals.

polyethylene: A thermoplastic insulation having excellent electrical properties, good chemical resistance (useful as jacketing), are good mechanical properties with the exception of temperature rating.

polymer: A high-molecular-weight compound whose structure can usually be represented by a repeated small unit.

polyphase circuits: ac circuits having two or more interrelated voltages, usually of equal amplitudes, periods, phase differences, etc.; if a neutral conductor exists, the voltages referenced to the neutral are equal in amplitude and phase; the most common version is that of three-phase, equal in amplitude with phases 120° apart. See Fig. P-7.

polyphase motor: An ac motor that is designed for either three- or two-phase operation. The two types are alike in construction, but the internal connections of the coils are different. Three-phase motors vary from fractional horsepower size to several thousand horsepower. These motors have a fairly constant speed characteristic and are made in designs giving a variety of torque characteristics. Some have a high starting torque; others have a low starting torque. Some are designed to draw a normal starting current; others draw a high

Fig. P-7: Three-phase cycle.

starting current. They are made for practically every standard voltage and frequency and are very often dual-voltage motors. Three-phase motors are used to drive machine tools, pumps, elevators, fans, cranes, hoists, blowers, and many other machines.

Two-phase motors are like three-phase motors in all respects, except for the number of groups and the connections of the groups. As in the three-phase motor, the number of groups is equal to the number of phases multiplied by the number of poles.

polypropylene: A thermoplastic insulation similar to polyethylene, but with slightly better properties.

polytetrafluoroethylene (PTFE): A thermally stable (−90 to +250°C) insulation having good electrical and physical properties even at high frequencies.

pop riveting tool: A tool that sets rivets without using a hammer, but a pre-drilled hole is needed to set the rivet.

porcelain: Ceramic china-like coating applied to steel surfaces.

portable: Designed to be movable from one place to another, not necessarily while in operation.

portable band saw: A hand-held power saw. The saw blade is a continuous loop and runs on two rubber tire type wheels. It is used to cut conduit, plastic pipe, or small metal angle irons.

positive: Connected to the positive terminal of a power supply.

positive charge: An object that has a majority of positive ions.

positive ion: An atom that is positively charged because of a missing electron.

positive plate: The plate of a storage cell from which the current flows to the negative plate during the process of discharging.

potential: The difference in voltage between two points of a circuit. Frequently, one is assumed to be ground (zero potential).

potential energy: Energy of a body or system with respect to the position of the body or the arrangement of the particles of the system.

potentiometer: An instrument for measuring an unknown voltage or potential difference by balancing it, wholly or in part, by a known potential difference produced by the flow of known currents in a network of circuits of known electrical constants.

potentiometer circuit: A circuit that contains variable resistors.

pothead: A terminator for high-voltage circuit conductor to keep moisture out of the insulation and to protect the cable end, along with providing a suitable stress relief cone for shielded-type conductors. See Fig. P-8.

power: 1) Work per unit of time. 2) The time rate of transferring energy; as an adjective, the word "power" is descriptive of the energy used to perform useful work; measurements: pound-feet per second, watts.

power, active: In a 3-phase symmetrical circuit: $p = \sqrt{3}$ VA cos ϕ; in a 1-phase, 2 wire circuit: p = VA cos ϕ.

Fig. P-8: Pothead used on a power pole to make the transition from overhead to underground lines.

power, apparent: The product of rms volts times rms amperes.

power consumption: The amount of energy used to operate an electrical device.

power element: Sensitive element of a temperature-operated control.

power factor: Correction coefficient for ac power necessary because of changing current and voltage values. The power factor of an ac circuit is a ratio of the apparent power compared to the true power. Power factor is important because utility companies charge industries large penalties for a poor power factor. The power factor of a circuit or system can be found by the equation:

$$\text{PF} = \frac{\textit{true power (W)}}{\textit{apparent power (VA)}}$$

power factor correction: Capacitance is used for power factor correction. When the capacitive current is at its peak positive value, the inductive current is at its peak negative value, and vice-versa.

Since these two currents are in direct opposition, one can be used to cancel the other. To determine the amount of capacitance for power factor correction, first calculate the power factor (see *power factor*). Find the reactive power (vars) using the equation:

$$\text{Vars} = \sqrt{VA^2 - W^2}$$

This reactive power in the circuit can be canceled by an equal amount of capacitive vars, found by the equation $X_C = E^2/\text{vars}$. Finally, the amount of capacitance, in farads, needed to produce the correction can be found using the following equation:

$$C = \frac{1}{2 \times \pi \times \text{frequency} \times X_C}$$

power-limited cable: Cable that is limited to lower voltages and power levels than regular line-voltage cable.

power loss (cable): Losses due to internal cable impedance, mainly I^2r; the losses cause heating.

power panelboard: An electrical cabinet placed in or against a wall and accessible only from the front (*NEC®*).

power pool: A group of power systems operating as an interconnected system.

powers (math): Is the number of times a number is multiplied by itself.

power source: The point from which electrical power originates.

power supply: A source from which independent system components derive their power.

P-P: Peak to Peak.

precast concrete: Concrete units (such as piles or vaults) cast away from the construction site and set in place.

precious metal: Gold, silver, or platinum.

precipitators: Devices that use static electricity to clean particles from the air.

premolded: A splice or termination manufactured of polymers, ready for field application.

pressure: 1) An energy impact on a unit area; force or thrust exerted on a surface. 2) Electromotive force commonly called *voltage.*

pressure motor control: A device that opens and closes an electrical circuit as pressures change.

pressure switch: A sensing device that can detect either the absence or presence of pressure; its function is to either open or close a set of contacts.

preventive maintenance: Scheduled maintenance of equipment, assembly lines, and motors to prevent costly unscheduled shutdowns.

prewired: Electrical equipment or light fixtures that are ready to hook up to the branch circuit without any additional wiring.

primary: Normally referring to the part of a device or equipment connected to the power supply circuit.

primary cell: A device for transforming chemical energy into electrical energy. It consists essentially of a container with a solution (electrolyte) and two plates of electrodes.

primary coil: The coil into which the original energy is introduced and which sets up magnetic lines of force to link with another coil in which energy is induced.

primary control: Device that directly controls operation of a heating system.

printed circuit: A board having interconnecting wiring printed on its surface and designed for mounting of electronic components.

process: Path of succession of states through which a system passes.

program, computer: The ordered listing of sequence of events designed to direct the computer to accomplish a task.

programmable thermostat: A thermostat that is capable of controlling temperature, time of operation, humidity, ventilation, filtration, and circulation of air as well as zone control.

propagation: The travel of waves through or along a medium.

property: An observable characteristic.

proportion: Two ratios that are equal to each other.

proportional: Where things are properly related to in size or other ways. A constant ratio.

proprietary supervising station (fire alarms): A fire station where personnel are in attendance at all times to monitor the fire alarm signals.

protected enclosure: Having all openings protected with screening, etc.

protector, circuit: An electrical device that will open an electrical circuit if excessive electrical conditions occur.

proton: The mass of the proton is $1.6726485 \times^{-27}$kg, or 1.007276470 u on the atomic mass scale. The positive charge of a proton has the same magnitude as the negative charge of an electron.

prototype: The first full-size working model.

proximity effect: The distortion of current density due to magnetic fields; increased by conductor diameter, close spacing, frequency, and magnetic materials such as steel conduit or beams.

PSAR (Preliminary Safety Analyses Report) (nuclear): Construction permit.

PSI: Pound force per Square Inch.

PT: Potential transformer.

P-type material: A crystalline structure that has holes in it, with the holes acting as carriers when voltage is applied.

pull box: A sheet metal box-like enclosure used in conduit runs, either single conduits or multiple conduits, to facilitate pulling in of cables from point to point in long runs, or to provide installation of conduit support bushings needed to support the weight of long riser cables, or to provide for turns in multiple-conduit runs.

pull-down: Localized reduction of conductor diameter by longitudinal stress.

pulling compound (lubricant): A substance applied to the surface of a cable to reduce the coefficient of friction during installation.

pulling eye: A device attached to a cable to facilitate field connection of pulling ropes.

pull-up torque: Torque that is required to bring a motor up to speed.

pulsating current: A direct current in which the value is not constant but the flow is in one direction.

pulsating dc: A dc current that returns to zero intermittently through the use of an ac rectifier.

pulsating function: A periodic function whose average value over a period is not zero.

pulse: A brief excursion of a quantity from normal.

pumped storage (hydro power): The storage of power by pumping a reservoir full of water during off-peak, then depleting the water to generate when needed.

puncture: Where breakdown occurs in an insulation.

purge: To clean.

push button: 1) A device that completes an electric circuit as long as a button is depressed. 2) A switch in which electrical contacts are closed by pushing one button and are opened by pushing another.

PVC (Polyvinyl Chloride): A thermoplastic insulation and jacket compound.

PWR (Pressurized Water Reactor) (nuclear power): A basic nuclear fission reaction in which water is used to transfer energy from the reactor; the water exchanges its heat with a secondary loop to produce steam for the turbine.

pyroconductivity: Electric conductivity that develops with changing temperature, and notably upon fusion, in solids that are practically non-conductive at atmospheric temperatures.

pyrometer: Thermometer that measures the radiation from a heated body.

Pythagorean theorem: The square of the hypotenuse of a right triangle equals the sum of the squares of the other two sides.

QA (Quality Assurance): All the planned and systematic actions to provide confidence that a structure, system, or component will perform satisfactorily.

QAP (nuclear): Quality Assurance Policy.

Q rating: The ratio of resistance to capacitance reactance in a capacitor. The ratio of inductance reactance to resistance in a conductor. A measure of quality of the component.

quadrant: The quarter of a circle, or of its circumference.

quadrilateral: A four-sided polygon consisting of four coplaner points that intersect only at their end points.

quadruplexed: Twisting of four conductors together.

qualified life (nuclear power): The period of time for which satisfactory performance can be demonstrated for a specific set of service conditions.

qualified person: A person familiar with construction, operation, and hazards.

quality: A number that expresses the ratio of reactance to resistance.

quarter bend: A bend through an arc of 90 degrees; as in a piece of conduit. Sometimes called *elbow*.

quartz: A popular bulb material because of its high lumens. High operating temperatures require good heat-dissipating characteristics of the lamp holder.

quick-: Prefix indicating a device that has a high contact speed independent of the operator; example: quick-make or quick-break. See description of *quick-break switch* for details of operation.

quick-break switch: Usually of the knife-blade type. The blade is made of two pieces. As the switch is pulled out, the first half of the blade is withdrawn, and as the throw increases, the second half is drawn out by the action of a spring attached to the first blade.

quiet switch: A snap switch that has a very quiet operation compared to the old type snap switch, which had a very loud operation. Many of the new quiet switches are now electronic.

quotation: A firm price given a buyer. For example, "the supplier quoted a price of $200 for a 200-amperes, three-phase disconnect switch" or "The electrical contractor's quotation for the electrical work was $17,500."

R

raceway: Any channel designed expressly for holding wire, cables, or bars and used solely for that purpose. See Fig. R-1.

Fig. R-1: Raceways attached to a strut support system.

rack (cable): A device to support cables.

radar: A radio detecting and ranging system.

radial feeder: A feeder connected to a single source.

radian (rad): A supplementary SI unit for plane angles; the plane angle with its vertex at the center of a circle that is subtended by an arc equal in length to the radius.

radiant energy: Energy traveling in the form of electromagnetic waves.

radiant heating: Heating system in which warm or hot surfaces are used to radiate heat into the space to be conditioned.

radiation: The process of emitting radiant energy in the form of waves or particles.

radiation, blackbody: Energy given off by an ideal radiating surface at any temperature.

radiation, nuclear: The release of particles and rays during disintegration or decay of an atom's nucleus; these rays cause ionization; they are: alpha particles, beta particles, gamma rays.

radius, bending: The radii around which cables are pulled.

radius, training: The radii to which cables are bent by hand positioning, not while the cables are under tension.

rail clamp: A device to connect cable to a track rail.

rainproof: So constructed, protected, or treated as to prevent rain from interfering with the successful operation of the apparatus under specified test conditions.

rainshield: An inverted funnel to increase the creepage over a stress cone.

raintight: So constructed or protected that exposure to a beating rain will not result in the entrance of water. See Fig. R-2.

Fig. R-2: Raintight motor control.

raked joint: Joint formed in brickwork by raking out some of the mortar an even distance from the face of the wall.

ram: Random access memory.

ratchet pipe threader: A manually operated tool that has a ratchet built into the handle. The handle will hold dies in sizes from 1/8 inch to 2 inches.

rated: Indicating the limits of operating characteristics for application under specified conditions.

rated life: The median life of a bulb.

rated-load current: The current when a hermetic-compressor is operating at the rated load, rated voltage, and rated frequency.

rate-of-rise detector: A device that responds to any rapid rise of temperature above a predetermined amount within a designated time period.

rating, temperature (cable): The highest conductor temperature attained in any part of the circuit during: a) normal operation, b) emergency overload, c) short circuit.

rating, voltage: The thickness of insulation necessary to confine voltage to a cable conductor after withstanding the rigors of cable installation and normal operating environment.

ratio: The comparison between two similar things. Expressed as a fraction.

RC time constant: Time = Resistance × Capacitance.

REA (Rural Electrification Administration): A federally supported program to provide electrical utilities in rural areas.

reactance: 1) The imaginary part of impedance. 2) The opposition to ac due to capacitance (X_C) and/or inductance (X_L).

reactor: A device to introduce capacitive or inductive reactance into a circuit.

reactor, nuclear: An assembly designed for a sustained nuclear chain reaction; the chain reaction occurs when the mass of the fuel reaches a critical value having enough free neutrons or enough heat to sustain fusion.

real time: The actual time during which a physical process transpires.

reamer: A finishing tool that finishes a circular hole more accurately than a drill. A tool for removing burrs on the inside of the mouth on metal conduit. See Fig. R-3.

receiver (fiber optics): A device that converts light to an electrical signal in a fiber-optic-link.

receptacle: A contact device installed at an outlet for the connection of an attachment plug and flexible cord to supply portable equipment. See Fig. D-14.

Fig. R-3: A pipe reamer is used to remove burrs from the inside of metal conduit.

reciprocating: Action in which the motion is back and forth in a straight line.

reciprocating saw: A special saber saw used where larger saws will not fit. Used to cut metal, wood, sheetrock, and conduit.

recoded wire: A white wire's ends are taped black to indicate that the wire carries power. Wires are taped green to indicate that they are ground wires. See Fig. R-4.

recognized: An acknowledgment of authority.

recognized component: An item to be used as a subcomponent and tested for safety by an independent testing laboratory.

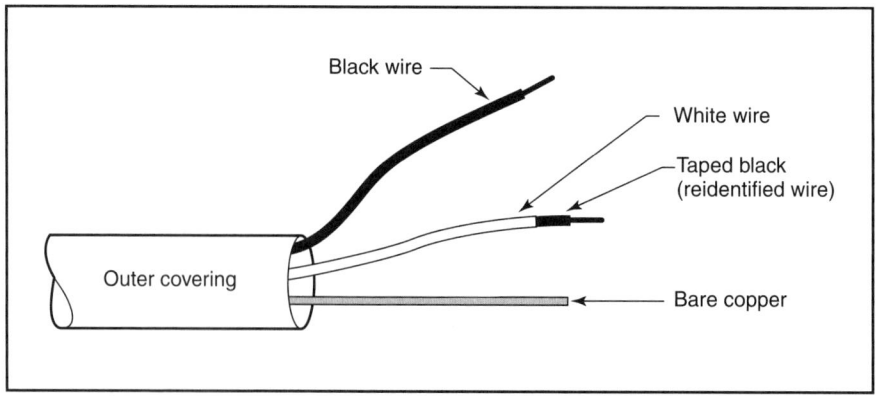

Fig. R-4: The color of recoded wire is changed from its original color.

recorder: A device that makes a permanent record, usually visual, of varying signals.

rectifiers: Devices used to change alternating current to direct current.

rectify: To change from ac to dc.

red-leg: See *high-leg*.

redraw: Drawing of wire through consecutive dies.

reduced size neutral: The neutral conductor may only be sized smaller than the phase conductor but must be sized for the maximum neutral current. *NEC® 220-22.*

reducing joint: A splice of two different size conductors.

reduction: The gain of electrons by a constituent of a chemical reaction.

redundancy: The use of auxiliary items to perform the same functions for the purpose of improving reliability and safety.

reel: A drum having flanges on the ends; reels are used for wire/cable storage.

reference node: A common connection in a circuit from which all voltages are measured. See Fig. R-5.

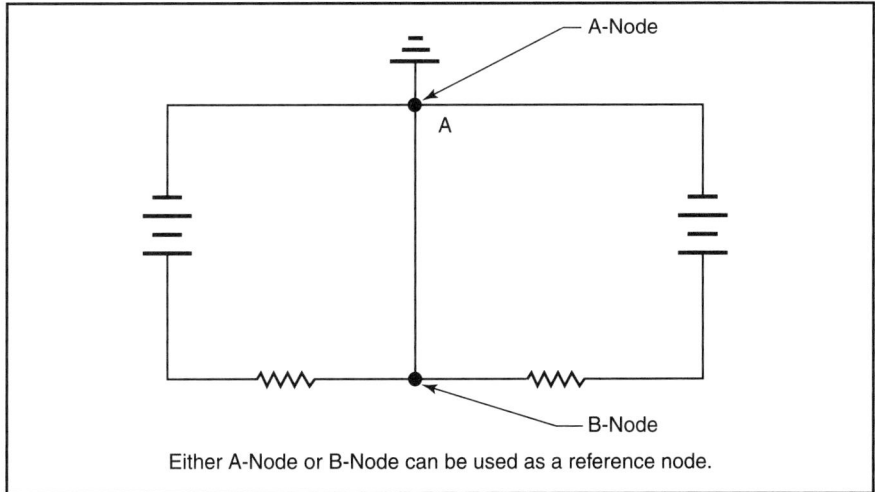

Fig. R-5: Reference node is a common connection in a circuit from which all voltages are measured.

reference symbols: Used for detailed drawings or for details in section drawings.

reflective insulation: Thin sheets of metal or foil on paper set in the exterior walls of a building to reflect radiant energy.

reflector: An incandescent bulb that has a coating on the inside, which allows the light to be directed to a certain area.

refrigerant: Substance used in refrigerating mechanisms to absorb heat in an evaporator coil and to release heat in a condenser as the substance goes from a gaseous state back to a liquid state.

register: Combination grille and damper assembly covering on an air opening or end of an air duct.

regulation: The maximum amount that a power supply output will change as a result of the specified change in line voltage, output load, temperature, or time.

regulator: A device that keeps voltage constant regardless of the current drain.

reinforced jacket: A cable jacket with reinforcing fiber between layers.

relative capacitance: The ratio of a material's capacitance to that of a vacuum of the same configuration; will vary with frequency and temperature.

relative humidity: Ratio of amount of water vapor present in air to greatest amount possible at same temperature.

relative value: The tolerance by which a product may vary from its designed value.

relay: A device designed to abruptly change a circuit because of a specified control input. See Fig. R-6 on page 227.

relay, overcurrent: A relay designed to open a circuit when current in excess of a particular setting flows through the sensor.

reliability: The probability that a device will function without failure over a specified time period or amount of usage.

relief valve: Safety device to permit the escape of steam or hot water subjected to excessive pressures or temperatures.

reluctance: Resistance to magnetic flux.

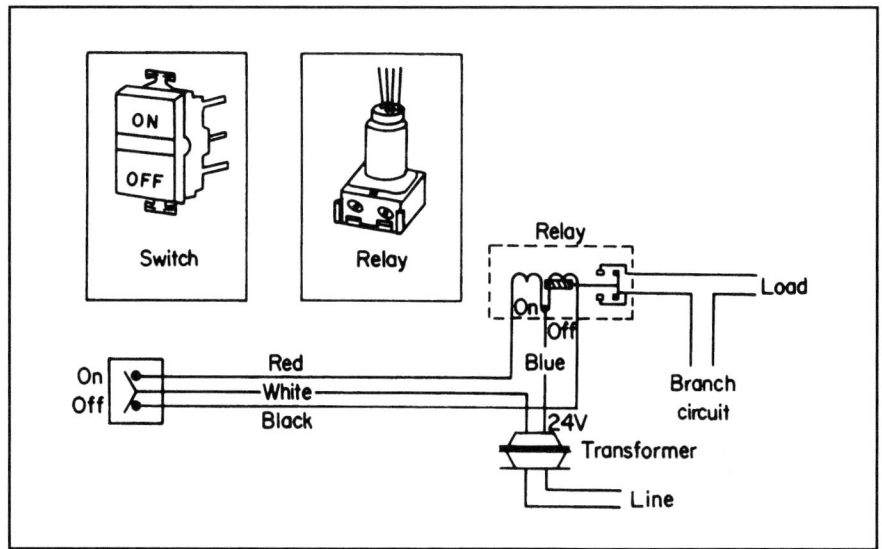

Fig. R-6: A relay circuit with the required components for proper funtioning.

remote-control circuits: The control of a circuit through relays, etc.

repeatability: The closeness of agreement among repeated measurements of the same variable under the same conditions.

reproducibility: The ability of a system or element to maintain its output or input over a relatively long period of time.

repulsion: A force that repels charged objects.

repulsion motor: Instantaneous polarity of rotary poles is the same as the adjacent stator poles forming a repulsion torque. Like poles repel. See Fig. R-7.

Fig. R-7: Repulsion motor (windings) work on the principle that like poles repel.

reservoir, thermal: A body to which and from which heat can be transferred indefinitely without a change in the temperature of the reservoir.

residual elements: Elements present in an alloy in small quantities, but not added intentionally.

residual stress: Stress present in a body that is free of external forces or thermal gradients.

resin: The polymeric base of all jacketing, insulating, etc., compounds, both rubber and plastic.

resistance: The opposition in a conductor to current; the real part of impedance.

resistance furnace: A furnace that heats by the flow current against ohmic resistance internal to the furnace.

resistance-start induction-run motor: A type of split-phase motor that uses the resistance in the start winding to cause a phase shift between the start winding current and the run winding current. See Fig. R-8.

resistance, thermal: The opposition to heat flow; for cables it is expressed by degrees centigrade per watt per foot of cable.

resistance welding: Welding by pressure and heat when the work piece's resistance in an electric circuit produces the heat.

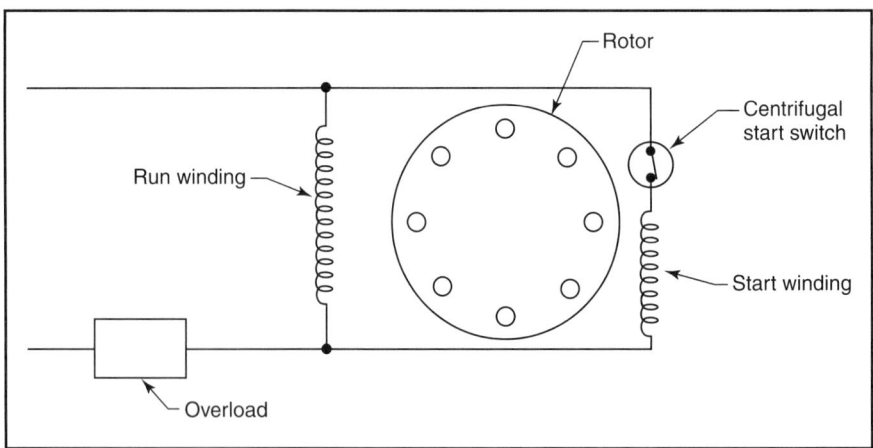

Fig. R-8: Resistance-start induction-run motor (windings).

resistive load: An electrical load that is void of any in-rush current.

resistivity: A material characteristic opposing the flow of energy through the material; expressed as a constant for each material; it is affected by temper, temperature, contamination, alloying, coating, etc.

resistor: A device whose primary purpose is to introduce resistance.

resistor, bleeder: 1) Used to drain current after a device is de-energized. 2) Improves voltage regulation. 3) Protects against voltage surges.

resolution: The degree to which nearly equal values of a quantity can be discriminated.

resolver: A device whose input and output is a vector quantity.

resonance: A condition reached in an electrical circuit when the inductive reactance neutralizes the capacitance reactance leaving ohmic resistance as the only opposition to the flow of current.

resonating: The maximizing or minimizing of the amplitude or other characteristics provided the maximum or minimum is of interest.

response: A quantitative expression of the output as a function of the input under conditions that must be explicitly stated.

restrike: A resumption of current between contacts during an opening operation after an interval of zero current of ¼ cycle at normal frequency or longer.

reverse lay: Reversing the direction of lay about every five feet during cabling of aerial cable to facilitate field connections.

reversible process: Can be reversed and leaves no change in system or surroundings.

revolving armature: A three-phase alternator's armature that has slip rings instead of a commutator.

RF: Radio Frequency: 10kGz to GHz.

RFI: Radio Frequency Interference.

rheology: The science of the flow and deformation of matter.

rheostat: A variable resistor that can be varied while energized, normally one used in a power circuit.

right-angle drill motor: A hand-held electric drill that has a right angle head making it very useful for drilling holes in wooden studs or in other areas that have limited working space.

right-hand rule: Point the forefingers of the right hand in the direction of the magnetic field. Point your center finger in the direction of the current flow. The thumb will indicate the direction of armature rotation. See Fig. R-9.

right triangle: Any triangle that has a 90° angle.

ring conductor: The telephone conductor that is connected to the negative terminal. See Fig. C-2.

ring-out: 1) A circular section of insulation or jacket. 2) The continuity testing of a conductor.

ripple: The ac component from a dc power supply arising from sources within the power supply.

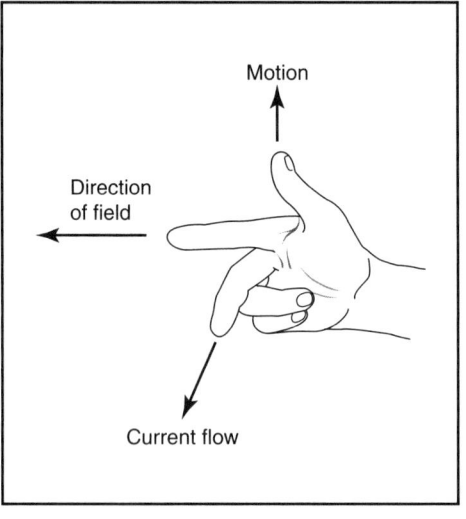

Fig. R-9: Right-hand rule. The forefinger points in the direction of the magnetic field. The center finger points in the direction of current flow. The thumb points in the direction of armature rotation.

rise (pipe bending): A measurement from the end of the pipe to the bottom of the bend.

riser: A vertical run of conductors in conduit or busway, for carrying electrical power from one level to another in a building.

riser cable: Electrical cable rated to run vertically in a building shaft from floor to floor.

riser diagram: Electrical drawing that provides an instant overview of the electric service, related feeders, and subpanels. See Fig. R-10, on page 231.

Fig. R-10: Typical riser diagram showing an overview of the building's electric service and related components.

riser valve: Device used to manually control flow of refrigerant in vertical piping.

rivet: A short, metal, bolt-like fastening, without threads, which is clinched by hammering. Rivets are designated by the shape of the beads as flat or panhead, buttonhead, countersink, mushroom, soap, or swollen neck.

RMS (Root-Mean-Square): The square root of the average of the square of the function taken through the period.

rock duster: A machine to distribute rock dust over coal to prevent dust explosions.

rocker arm switch: A switch that operates with a rocker arm instead of a toggle.

rod: The shape of solidified metal convenient for wire drawing, usually ⁵⁄₁₆ inch or larger.

roller bearing: A bearing made of hardened-steel rollers instead of the round steel balls used in ball bearings.

rolling: Reducing the cross-sectional area of metal stock or otherwise shaping metal products using rotating rolls.

ROM: Read Only Memory.

Romex: General Cable's trade name for type NM cable; but it is used generically by electrical workers to refer to any nonmetallic sheathed cable. See Fig. R-11. Also see Fig. R-12 on page 233.

rope job: An installation of non-metallic sheathed cable.

roster: A tabulated schedule or list of names.

rotary: Turning on its axis, like a wheel.

rotary blower: An encased rotating fan such as that used for forced draft in furnaces.

Fig. R-11: Romex or type NM cable is mostly used for residential and small commerical wiring systems.

Fig. R-12: *NEC*® rules summarizing the installation of type NM (Romex) cable.

rotary converter: A single machine connected to an ac circuit that delivers dc or vice versa.

rotary dimmer switch: A rotary knob that allows the operator to select a wide range of light intensity.

rotation joint: The part of the satellite dish that allows the dish to be rotated.

rotor: Rotating part of a mechanism.

roughing in: The first stage of an electrical installation, when the raceway, cable, wires, boxes and other equipment are installed; that electrical work which must be done before any finishing or cover-up phases of building construction can be undertaken.

rough inspection: The first inspection made of an electrical installation after the conductors, boxes, and other equipment have been installed in a building under construction.

round off: To delete the least significant digits of a numeral and adjust the remaining by given rules.

RPM: Revolutions Per Minute.

RS232: An EIA standard for the transmission of data over a twisted-wire pair.

RS422: An EIA standard for allowing data transmission to be received over multiple locations.

rubber, chlorosulfonated polyethylene (CP): A synthetic rubber insulation and jacket compound developed by DuPont as Hypalon®.

rubber, ethylene propylene: A synthetic rubber insulation having excellent electrical properties.

running board: 1) A device to permit stringing more than one conductor simultaneously. 2) Board used as protection for Romex cable in a residential attic or basement.

run winding: One of the two windings in a split-phase motor. It is the winding that does not drop out after the motor has started.

rupture stress: The unit stress at the time of failure.

S

sacrificial protection: Prevention of corrosion by coupling a metal to an electrochemically more active metal that is sacrificed.

saddle bend: A bend designed to go over an object and back down to the original plane, creating a hump in the pipe.

SAE: Society of Automotive Engineers.

safety conductor: A safety sling used during overhead line construction.

safety control: A device that will stop the refrigerating unit if unsafe pressures and/or temperatures are reached.

safety factor: The ratio of the maximum stress that something can withstand to the estimated stress that it can withstand.

safety motor control: Electrical device used to open a circuit if the temperature, pressure, and/or the current flow exceed safe conditions.

safety plug: Device that will release the contents of a container above normal pressure conditions and before rupture pressures are reached.

safety switch: A knife-blade switch enclosed in a metal box and operated externally.

sag: The difference in elevation of a suspended conductor.

sag, apparent: Sag between two points at 60°F and no wind.

sag, final: Sag under specified conditions after the conductor has been externally loaded, then the load removed.

sag, initial: Sag prior to external loading.

sag, maximum: Sag at midpoint between two supports.

sag section: Conductor between two snubs.

sag snub: Where a conductor is held fixed and the other end moved to adjust sag.

sag, total: Sag under ice loading.

sampling: A small quantity taken as a sample for inspection or analysis.

saturation: The condition existing in a circuit when an increase in the driving signal does not produce any further change in the resultant effect. A magnetic material is said to be saturated when, upon increasing the ampere turns, no increase in the number of magnetic lines of force is obtained.

SBCCI: Southern Building Codes Congress International, Inc.

scaffolding: A temporary platform that is used in the construction industry.

scalar: A quantity (as mass or time) that has a magnitude described by a real number and no direction.

scan: To examine sequentially, part by part.

scattering: A fragmentation of light caused by striking unwanted obstacles resulting in a loss in the light-carrying capacity of fibers.

scavenger pump: Mechanism used to remove fluid from sump or containers.

schedules: Data tables and charts for installation of materials, products, and equipment found on blueprints, specification sheets, and schematics.

schematic (electrical): A one-line diagram that shows the electrical sequence of the components but does not show the physical location or appearance of components.

scintillation: The optical photons emitted as a result of the incidence of ionization radiation.

scope: Slang for oscilloscope.

Scotch Lok®: A brand name for a wire nut.

SCR: Silicon Controlled Rectifier. A semiconductor that has to have a pulse applied to the gate before it will conduct.

scram (nuclear): The rapid shutdown of a nuclear reactor.

screen pack: A metal screen used for straining.

screwdriver: 1) A blade screwdriver is designed to fit a straight slot in a screw. 2) A Phillips head screwdriver is designed to fit a cross-shaped slot in a screw.

SE: Service Entrance.

sealed: Preventing entrance.

sealed motor compressor: A mechanical compressor consisting of a compressor and a motor, both of which are enclosed in the same sealed housing, with no external shaft or shaft seals, and with the motor operating in the refrigerant atmosphere.

sealing compound: The material poured into an electrical fitting to seal and minimize the passage of vapors. A seal-off used in hazardous locations and filled with sealing compound is shown in Fig. S-1.

secondary: The second circuit of a device or equipment, which is not normally connected to the supply circuit. In a step-down transformer, the high-voltage side is called *primary* while the low-voltage side is called *secondary*.

Fig. S-1: Seal-off with sealing compound.

secondary winding: The output winding of a transformer.

section view: A cutaway view of a building or part.

security lighting: Any lighting that is provided for security or safety reasons.

Seebeck Effect: The generation of a voltage by a temperature difference between the junctions in a circuit composed of two homogeneous electrical conductors of dissimilar composition; or in a nonhomogeneous conductor the voltage produced by a temperature gradient in a nonhomogeneous region.

segment bend: Complex bend made by a series of short bends.

selective coded system: A fire alarm system that identifies the location of a fire in a building and that also contacts firefighting personnel.

selective coordination: An overcurrent protection system that isolates a fault in just a portion of the complete circuit and does not shut down the whole circuit.

selectivity: A circuit with the capability to separate voltages that have different frequencies.

selector switch: A rotator switch that has several positions for different circuit possibilities.

selenium: An element that has the capability of changing its conductivity as light intensity increases or decreases.

self-excitation: Direct current obtained from the brushes of a dc generator to provide current for its electromagnetic field. In alternators the term refers to a dc generator built on the alternator shaft. Used to provide direct current for the alternator field.

self-excited alternator: An ac generator that also produces a direct current for magnetizing its main fields.

self inductance: Magnetic field induced in the conductor carrying the current.

semiautomatic starter: An electrical starter in which part of the starting operation is not automatic.

semiconductor: A material that has electrical properties of current flow between a conductor and an insulator.

semimagnetic control: An electrical controller whose electromagnets control only part of the controller.

sensible heat: Heat that causes a change in temperature of a substance.

sensing device: A device that senses a change in the device that it is monitoring and alerts the device to act.

sensor: A material or device that goes through a physical change or an electronic characteristic change as conditions change.

separable insulated connector: An insulated device to facilitate power cable connections and separations.

separately excited: A machine that gets the current needed for the excitation of its field from some outside source.

separator: Material used to maintain physical spacing between elements in cables, such as a layer of tape to prevent jacket sticking to individual conductors.

sequence controls: Devices that act in series or in time order.

series: When two or more electrical components are connected so that the current feeding one must pass through the others, they are said to be connected in series (one after another, as in a string). See Fig. S-2. The following four rules state the condition that exists in a series circuit.

- The current is the same in all parts of a series circuit.
- The total resistance in a series circuit is equal to the sum of the individual resistances.
- The total voltage applied to a series circuit divides between the resistors in direct proportion to their resistance.

Fig. S-2: A series circuit consisting of a voltage source (battery) and four lamps connected in series by means of wire conductors.

- The sum of the voltage crops across all the resistors in a series resistive circuit is equal to the applied (source) voltage.

series-aiding: Two or more voltage sources brought together in a series circuit to increase total voltage.

series load: A circuit's load connected in series with the power supply.

series parallel circuit: A circuit made up of two or more simple parallel circuits all joined in series.

series power source: The source from which a series circuit gets its power; power supply, battery, alternator, etc.

series-rated system: An electrical system that incorporates several overcurrent protection devices that work together to limit the possibility of a fault current.

series-wound motor: A dc motor with the armature and field connected in series. Used on machines where variable loads occur. The speed varies with the load.

service: The equipment used to connect to the conductors runs from the utility line, including metering, switching and protective devices; also the electric power delivered to the premises, rated in voltage and amperes, such as a "100-amp, 120/240-volt, three-wire service." See Fig. S-3, on page 241.

service ampacity: The total amp rating of an electrical service.

service cable: The service conductors made up in the form of a cable.

service conductors: The supply conductors that extend from the street main or transformers to the service equipment of the premises being supplied.

service drop: Run of cables from the power company's aerial power lines to the point of connection to a customer's premises.

service entrance: 1) The electrical cable and devices that connect the utility transformer to the building metering device. 2) The point at which power is supplied to a building. See Fig. S-3, page 241.

service equipment: The necessary equipment, usually consisting of a circuit breaker or switch and fuses and their accessories, located near the point of entrance of supply conductors to a building and intended to constitute the main control and cutoff means for the supply to the building.

service factor: The allowable overload for a motor.

Fig. S-3: Major parts of an electric service.

service lateral: The underground service conductors between the street main, including any risers at a pole or other structure or from transformers, and the first point of connection to the service-entrance conductors in a terminal box, meter, or other enclosure with adequate space, inside or outside the building wall. Where there is no terminal box, meter, or other enclosure with adequate space, the point of connection is the entrance point of the service conductors into the building.

service panel: A metal box that holds the fuses, circuit breakers, neutral bus bar, GFIs, and any other devices that protect the circuit.

service point: The electrical connection at the point where the utility company's responsibility ends and the building owner's responsibility begins.

service raceway: The Rigid Metal Conduit, Electrical Metallic Tubing, or other raceway that encloses the service entrance conductors. See Fig. S-4 on page 243.

service valve: A device, attached to a refrigeration system, that provides an opening for gauges and/or charging lines.

serving: A layer of helically applied material.

servomechanism: A feedback control system in which at least one of the system signals represents mechanical motion.

setting (of circuit breaker): The value of the current at which the circuit breaker is set to trip.

seven-heat control: A control that is used for a cooking surface that has seven different settings.

shaded-pole induction motor: A motor that does not use outside help to get started and operates like a rotating magnet field. The motor has a low starting torque.

shaded-pole motor: A small ac motor that utilizes a shaded coil; used for light-start loads; has no brushes or commutator. See Fig. S-5 on page 243.

shading loop: A band connected to a magnetic pole to oppose a change in magnetic flux.

shaft furnace: A furnace used for pouring wire bars from continuous melting of cathodes.

Fig. S-4: Rigid Metal Conduit used as a service mask for a through-roof application.

shall: Mandatory requirement of the National Electrical Code.

shaving: Removing about 0.001 inch of metal surface.

shear: The lateral displacement in a body due to an external force causing sliding action.

sheath: A metallic close-fitting protective covering.

Fig. S-5: A schematic of a shaded-pole motor, which has a shaded coil and is used for light loads.

shell: The same as orbit. Pertaining to the electron paths that surround the atom.

shield: The conducting barrier against electromagnetic fields.

shield, braid: A shield of interwoven small wires.

shield, insulation: An electrically conducting layer to provide a smooth surface in intimate contact with the insulation outer surface; used to eliminate electrostatic charges external to the shield, and to provide a fixed known path to ground.

shield, tape: The insulation shielding system whose current carrying component is thin metallic tapes, now normally used in conjunction with a conducting layer of tapes or extruded polymer.

shim: Thin piece of material used to bring members to an even or level bearing.

shore feeder: From ship to shore feeder.

shore hardness: A measure of the hardness of a plastic.

short-circuit: An often unintended low-resistance path through which current flows around, rather than through, a component or circuit.

short cycling: Refrigerating system that starts and stops more frequently than it should.

short time (nuclear power): Operation for 2 hours out of a 24-hour period.

short-time overload rating (nuclear): The limiting overload current that one third (must be at least three) of the conductors in an assembly through a penetration can carry, with all other conductors fully loaded.

should: An *NEC*® term that means recommended but not required.

shrinkable tubing: A tubing that may be reduced in size by applying heat or solvents.

shrink chart: A chart that gives the cotangent of the bend degree. This will show the shrink value. Example: 40° × cotangent of 40° = shrink.

shroud: Housing over a condenser or evaporator.

shunt: A device having appreciable resistance or impedance connected in parallel across other devices or apparatus to divert some of the current; appreciable voltage exists across the shunt and appreciable current may exist in it.

shunt generator: A generator that has its shunt field connected in parallel to the armature.

shunt-wound motor: Used when the motor speed must be constant, irrespective of variation in load.

side prong base: A lamp base having a side prong that fits into a sleeve that has a groove.

sidewall load: The normal force exerted on a cable under tension at a bend; quite often called sidewall pressure.

side-wired receptacle: The terminal screws are mounted on the side of the receptacle.

side-wired switch: The terminal screws are mounted on the side of the switch.

signal: A detectable physical quantity or impulse (as a voltage, current, or magnetic field strength) by which messages or information can be transmitted.

signal circuit: Any electrical circuit supplying energy to an appliance that gives a recognizable signal. A simple residential doorbell circuit is shown in Fig. S-6, on page 246.

signal generator: A test instrument that produces low voltage for the purpose of testing electronic equipment.

significant digits (electrical math): The digits in a number that are confirmed plus the first digit that is not certain.

silent switch: See *quiet switch*.

silica gel: Chemical compound used as a drier.

silicon controlled rectifier (SCR): Electronic semiconductor that contains silicon.

sill: Horizontal timber forming the lowest member of a wood frame house; lowest member of a window frame.

similar poles: When two magnetic poles repel each other, they are magnetically similar or like.

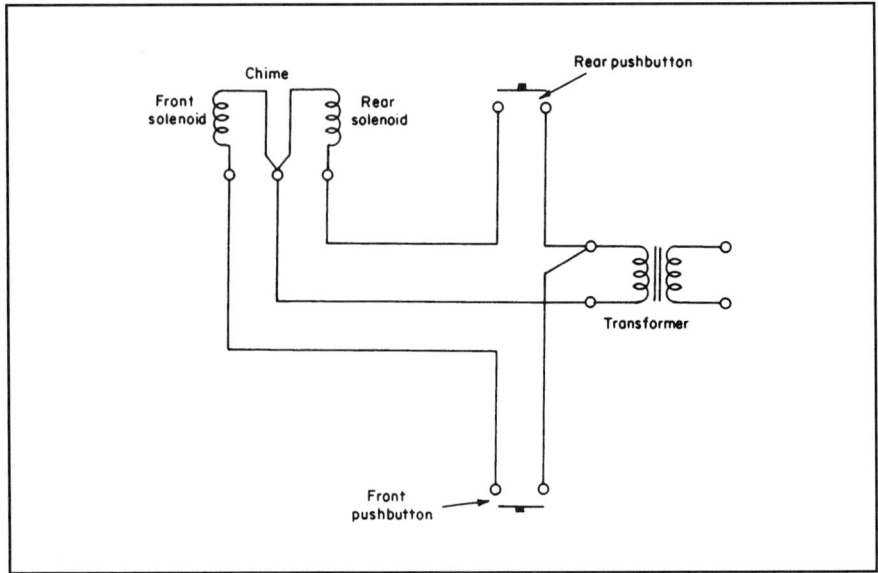

Fig. S-6: A wiring diagram showing a typical two-note chime controlled at two different locations.

sine wave, ac: Wave form of single frequency alternating current; wave whose displacement is the sine of the angle proportional to time or distance. See Fig. S-7.

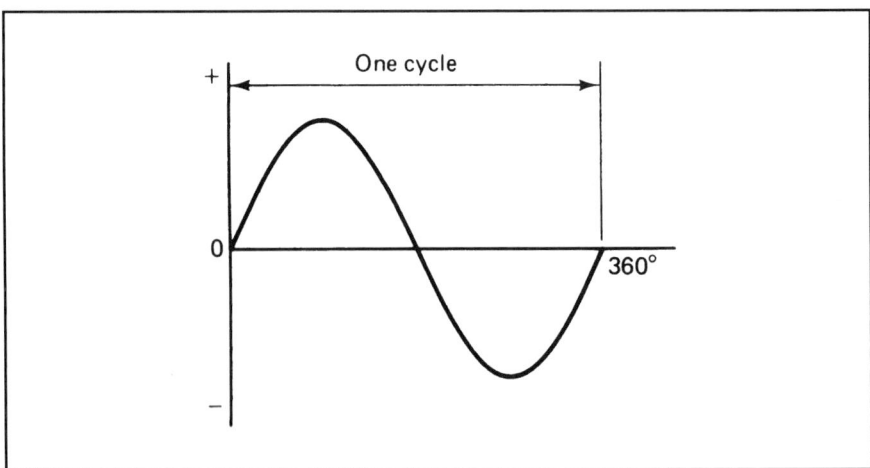

Fig. S-7: Typical sine wave.

single-phase circuit: An ac circuit having one source voltage supplied over two conductors.

single-phase motor: Electric motor that operates on single-phase alternating current.

single-phasing: The abnormal operation of a three-phase machine when its supply is changed by accidental opening of one conductor.

single-pole switch: A switch that opens and closes only one side of a circuit.

sintering: Forming articles from fusible powders by pressing the powder just under its melting point.

site plan: A plan that shows the property boundaries and the position of the building in relation to the boundaries. Generally, the site plan also shows any existing utilities and future outside utilities and bench mark locations.

skin effect: The tendency of current to crowd toward the outer surface of a conductor; increases with conductor diameter and frequency.

skinning: Removing insulation from electrical conductors before making splices or connections.

slip: The difference between the speed of a rotating magnetic field and the speed of its rotor.

slippercoat: A surface lubricant factor applied to a cable to facilitate pulling, and to prevent jacket sticking.

slip rings: The means by which the current is conducted to a revolving electrical circuit. See Fig. S-8, on page 248.

sliver: A defect consisting of a very thin elongated piece of metal attached by only one end to the parent metal into whose surface it has been rolled.

slot: A channel opening in the stator or rotor of a rotating machine for ventilation and the insertion of windings.

smoke detector: A device that senses the presence of visible and non-visible smoke particles and gives off a warning alarm.

snake: See *fish tape*.

Fig. S-8: Slip rings on a wound-rotor motor schematic.

snap action: A spring-loaded contact that opens and closes.

snap switch: 1) An older type of switch that had a knife blade contact and a very strong spring. 2) A general-use toggle switch.

SNI: Standard Network Interface.

SNR: Signal to Noise Ratio.

soap: Slang for cable pulling compound.

socket: The receptacle into which the threaded portion of an incandescent lamp or plug is fitted.

soffit: Underside of a stair, arch, or cornice.

soft drawn: 1) A relative measure of the tensile strength of a conductor. 2) Wire that has been annealed to remove the effects of cold working. 3) Drawn to a low tensile.

solar cell: The direct conversion of electromagnetic radiation into electricity; certain combinations of transparent conducting

Fig. S-9: Solar panel consisting of a group of solar cells facing the sun.

films separated by thin layers of semiconducting materials. See Fig. S-9.

solar heat: Heat from visible and invisible energy waves from the sun.

solder: To braze with tin alloy.

solder pot: 1) When low- and sharp-melting-point metals reach their melting point. 2) A cast iron pot that sits on top of a propane burner and melts the solder to a liquid state. A smaller cast iron pot with a handle is used to apply the solder to the conductors for a solder connection.

solenoid: Electric conductor wound as a helix with a small pitch; coil.

solenoid valve: A valve that has a plunger and an electromagnetic solenoid. When the solenoid is energized or de-energized, the plunger opens or closes.

solidly grounded: No intentional impedance in the grounding circuit.

solid state: A device, circuit, or system that does not depend upon physical movement of solids, liquids, gases, or plasma.

solid-state ignition: A device that replaces a continuously burning pilot light with an electrical ignitor that lights the pilot only when heat is called for.

solid type PI cable: A pressure cable without constant pressure controls.

solid wire: Wire that is made in one piece and not in strands.

soluble oil: Specially prepared oil whose water emulsion is used as a curing grinding or drawing lubricant.

solution: 1) Homogenous mixture of two or more components. 2) The solving of a problem.

sones: Pertains to loudness. One sone is equal to the loudness of a sound of one kilohertz at 40 decibels above the threshold of a listener's hearing.

source: The point from which a circuit receives its electrical power or signal.

SP: Single Pole.

space heater: A heater for occupied spaces.

spacer plate: A plate that connects to an electrical junction box to increase the depth and square inches required to make additional space for wires and splices.

span: A conductor between two consecutive supports.

spark: A brilliantly luminous flow of electricity of short duration that characterizes an electrical breakdown.

spark gap: Any short air space between two conductors.

spark test: A voltage withstand test on a cable while in production with the cable moving; a simple way to test long lengths of cable.

SPDT: Single-Pole Double-Throw.

special circuits: Circuits other than lighting and power circuits that have limitations and/or conditions.

special-purpose motor: A motor so designed that it does not fall into the category of general-purpose motor. A NEMA standard.

specific gravity: The weight of a substance compared with an equal volume of water.

specific heat: Ratio of the quantity of heat required to raise the temperature of a body one degree to that required to raise the temperature of an equal mass of water one degree.

specs: Abbreviation for the word "specifications," which is the written precise description of the scope and details of an electrical installation and the equipment to be used in the system. A sample specification is shown in Fig. S-10, on page 252.

spectrum: The distribution of the amplitude (and sometimes phase) of the components of the wave as a function of frequency.

speed control switch: A switch that controls the speed of small motors by the use of several speed settings.

speed of recovery: The length of time it takes to bring a fluid back up to a desired temperature.

spike: A pulse having great magnitude.

splice: The electrical and mechanical connection between two pieces of cable. See *Western Union splice*, *fixture splice*, etc.

splice tube: The movable section of vulcanizing tube at the extruder.

splicing rope: Tying two ropes together by braiding or splicing.

split-circuit: A device that is supplied by two different circuits.

split-circuit receptacle: A duplex receptacle that has one half supplied by one circuit and the other half by a second circuit.

split fitting: A conduit fitting that may be installed after the wires have been installed.

split-phase motor: Motor with two stator windings. Winding in use while starting is disconnected by a centrifugal switch after the motor attains speed, then the motor operates on the other winding.

split-receptacle: A duplex receptacle that has one half controlled by a switch and whose other half is hot at all times.

split system: Refrigeration or air conditioning installation that places the condensing unit outside or remote from the evaporator. It is also applicable to heat pump installations.

DIVISION 16-ELECTRICAL

Section 16A-General Provisions

1. The "Instruction to Bidders," "General Conditions," and "General Requirements" of the architectural specifications govern work under this Section.

2. It is understood and agreed that the Electrical Contractor has, by careful examination of the Plans and Specifications, and the site where appropriate, satisfied himself as to the nature and location of the work, and all conditions which must be met in order to carry out the work under this Section of the Contract.

3. Scope of the Work

a. The scope of the work consists of the furnishing and installing of complete electrical systems--exterior and interior--including miscellaneous systems. The Electrical Contractor shall provide all supervision, labor, materials, equipment, machinery, and any and all other items necessary to complete the systems. The Electrical Contractor shall note that all items of equipment are specified in the singular; however, the Contractor shall provide and install the number of items of equipment as indicated on the drawings and as required for complete systems.

b. It is the intention of the Specifications and Drawings to call for finished work, tested, and ready for operation.

c. Any apparatus, appliance, material, or work not shown on drawings but mentioned in the specifications or vice versa, or any incidental accessories necessary to make the work complete and perfect in all respects and ready for operation, even if not particularly specified, shall be furnished, delivered, and installed by the Contractor without additional expense to the Owner.

d. Minor details not usually shown or specified, but necessary for proper installation and operation, shall

Fig. S-10: A sample page from an electrical specification.

split-wire: A way of wiring a duplex receptacle outlet with a 3-wire, 120/240 volt single-phase circuit so that one hot leg and the grounded conductor (neutral) feed one of the receptacle outlets, and the other hot leg and grounded conductor (neutral) feed the other receptacle outlet. This gives the capacity of two separate circuits to one duplex receptacle.

spray cooling: Method of refrigerating by spraying refrigerant inside the evaporator or by spraying refrigerated water.

springback: The number of degrees that a pipe tends to straighten when releasing the pipe benders after the pipe is bent.

SPST: Single-Pole Single-Throw.

spurious response: Any response other than the desired response of an electric transducer or device.

square of a number: A number multiplied by itself. Example: $5^2 = 25$.

square root: The inverse of raising to the power. Example: The square root of $25 = 5$.

squirrel-cage motor: An induction motor having the primary winding (usually the stator) connected to the power and a current is induced in the secondary cage winding (usually the rotor).

SSR: Solid State Relay.

stability factor: Percent change in dissipation factor with respect to time.

stack: Any vertical line of soil, waste, or vent piping.

standard conditions: Temperature of 68 degrees Fahrenheit, pressure of 29.92 inches of mercury, and relative humidity of 30 percent.

standard deviation: 1) A measure of data from the average. 2) The root mean square of the individual deviations from the average.

standard fuse: A regular screw-in type of fuse.

standard plug: A long plug whose line cord attaches to the screw terminals.

standard reference position: The nonoperated or de-energized condition.

standing wave: A wave in which, for any component of the field, the ratio of its instantaneous value at one point to that at any other point does not vary with time.

standoff: An insulated support.

star connection: Three-phase generators and transformers have three coils that may be connected in star, Y, or delta. When one terminal of each coil is connected together and the other three terminals are brought out separately, the connection is called *star* or *Y*.

starter: 1) An electric controller for accelerating a motor from rest to normal speed and for stopping the motor. 2) A device used to start an electric discharge lamp. See Fig. S-11.

starting relay: An electrical device that connects and/or disconnects the starting winding of an electric motor.

starting torque: The turning effort produced by a motor upon its shaft through the electromagnetic effect at the initial flow of current.

Fig. S-11: Simple motor starter.

starting winding: Winding in an electric motor used only during the brief period when the motor is starting.

start up: The time between the installation of equipment and the achievement of full running operation.

static: Interference caused by electrical disturbances in the atmosphere.

static control: Solid-state devices in which there are no moving parts.

stator: The portion of a rotating machine that includes and supports the stationary active parts.

steady state: A characteristic exhibiting only negligible change over a long period of time.

stealer transistor: A device that steals the base current away from another transistor, not allowing it to turn on.

steam: Water in vapor state.

steam heating: Heating system in which steam from a boiler is conducted to radiators in a space to be heated.

step-down transformer: The secondary coil has fewer turns of wire than the primary coil. Therefore the output voltage will be less than the input voltage. See Fig. S-12, on page 256.

step-up transformer: The secondary coil has more turns of wire than the primary coil. Therefore the output voltage will be greater than the input voltage. See Fig. S-12, on page 256.

steradian: The supplemental SI unit for solid angle: the three dimensional angle having its vertex at the center of a sphere and including the area of the spherical surface equal to that of a square with sides equal in length to the radius of the sphere.

storage cell: One of the sections of a storage battery.

strain: 1) A change in characteristic resulting from external forces. 2) To screen foreign materials from a substance.

strand: A group of wires, usually twisted or braided.

strand, annular: A concentric conductor over a core; used for large conductors (1000 kcmil @ 60 hertz) to make use of skin effect; core may be of rope, or twisted I-beam.

strand, bunch: A substrand for a rope-lay conductor; the wires in the substrand are stranded simultaneously with the same direction; bunched conductors flex easily and with little stress.

strand, class: A system to indicate the type of stranding; the postscripts are alpha.

strand, combination: A concentric strand having the outlet layer of different size; done to provide smoother outer surface; wires are sized with +5% tolerance from nominal.

strand, compact: A concentric stranding made to a specified diameter of 8%–10% less than standard by using smaller than normal closing die, and for larger sizes, preshaping the strands for the outer layer(s).

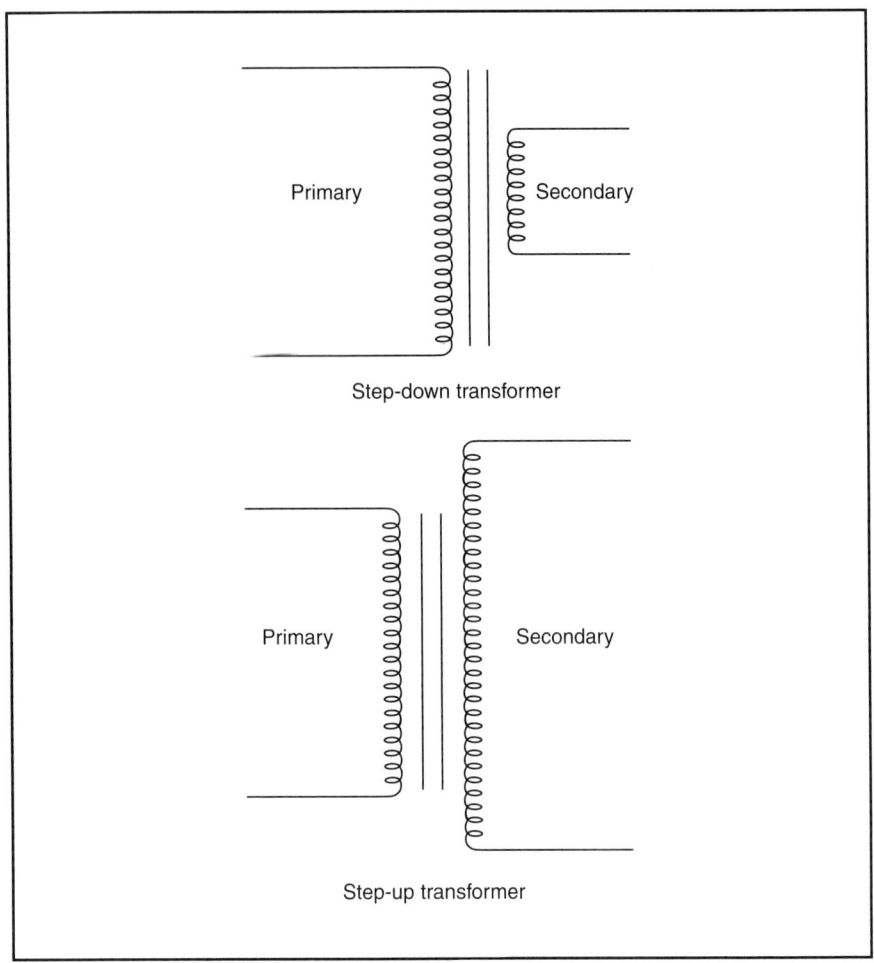

Fig. S-12: Step-down and step-up transformers. A step-down transformer secondary has fewer turns than the primary. A step-up transformer secondary has more turns than the primary.

strand, compressed: The making of a tight stranded conductor by using a small closing die.

strand, concentric: Having a core surrounded by one or more layers of helically laid wires each of one size, each layer increased by six.

strand, herringbone lay: Adjacent bunches having opposite direction of lay in a layer of a rope-lay cable.

strand, nonspecular: One having a treated surface to reduce light reflection.

strand, regular lay: Rope stranding having left-hand lay within the substrands and right-hand lay for the conductor.

strand, reverse-lay: A stranding having alternate direction of lay for each layer.

strand, rope-lay: A conductor having a lay-up of substrands; substrand groups are bunched or concentric.

strand, sector: A stranded conductor formed into sectors of a circle to reduce the overall diameter of a cable.

strand, segmental: One having sectors of the stranded conductor formed and insulated one from the other, operated in parallel; used to reduce ac resistance in single conductor cables.

strand, unilay (unidirectional): Stranding having the same direction of lay for all layers; used to reduce diameter, but is more prone to birdcaging.

stratification of air: Condition in which there is little or no air movement in the room; air lies in temperature layers.

stress: 1) An internal force set up within a body to resist or hold it in equilibrium. 2) The externally applied forces.

stress-relief cone: Mechanical element to relieve the electrical stress at a shield cable termination; used above 2kV.

striking: The process of establishing an arc or a spark.

striking distance: The effective distance between two conductors separated by an insulating fluid such as air.

stringers: Members supporting the treads and risers of a stair.

strip: To remove insulation or jacket.

strip cooler: A device to cool strips of compound.

strut: A compression member other than a column or pedestal.

stub ups (pipebending): Another name for rise.

stud locator: A magnetic instrument that detects nails in studs and joists.

studs: Vertically set skeleton members of a partition or wall to which the wallboard is nailed.

sub-panel: A panelboard in a residential system that is fed from the service panel; or any panel in any system that is fed from another, or main, panel supplied by a circuit from another panel.

substation: An assembly of devices and apparatus to monitor, control, transform, or modify electrical power. See Fig. S-13 on page 259.

subtractive polarity: When the primary and secondary windings of a transformer are connected in series, the output voltage is less than that of the winding that has the greater voltage.

superconductors: Materials whose resistance and magnetic permeability are infinitesimal at absolute zero (273°C).

superposition: A method used to find the current in any branch circuit that contains more than one power source.

superposition theorem: Current can be determined in any multi-power source by finding the current produced in any branch by each of the individual power sources.

supervised circuit: A closed circuit having a current responsive device to indicate a break or ground.

supplemental ground: Electrodes that are permitted to enhance the equipment grounding conductors.

surface-mounted fixture: Light fixtures that are designed to be mounted on the ceiling surface.

surge: 1) A sudden increase in voltage and current. 2) Transient condition.

surge arrestor: A device used on electrical circuits to detect and check sudden increases in voltage and current.

surge suppressor: A device that prevents voltage spikes or surges.

surround lighting: Lighting that is generally used for exterior lighting for safety and nighttime activities.

suspended ceiling fixture: A fluorescent light fixture that is designed to fit on the t-bars of a suspended ceiling grid. Also called a lay-in fixture.

Fig. S-13: A substation is used to monitor, control, and transform transmission power.

swag fixture: A light fixture suspended from the ceiling by a decorative chain.

switch: A device for opening and closing or for changing the connection of a circuit.

switch, ac general-use snap: A general-use snap switch suitable only for use on alternating current circuits and for controlling the following:

- Resistive and inductive loads (including electric discharge lamps) not exceeding the ampere rating at the voltage involved.
- Tungsten-filament lamp loads not exceeding the ampere rating of the switches at the rated voltage.
- Motor loads not exceeding 80 percent of the ampere rating of the switches at the rated voltage.

switch, ac-dc general-use snap: A type of general-use snap switch suitable for use on either direct or alternating current circuits and for controlling the following:

- Resistive loads not exceeding the ampere rating at the voltage involved.
- Inductive loads not exceeding one-half the ampere rating at the voltage involved, except that switches having a marked horsepower rating are suitable for controlling motors not exceeding the horsepower rating of the switch at the voltage involved.
- Tungsten-filament lamp loads not exceeding the ampere rating at 125 volts, when marked with the letter T.

switchboard: A large single panel, frame, or assembly of panels having switches, overcurrent, and other protective devices, buses, and usually instruments mounted on the face or back or both. Switchboards are generally accessible from the rear and from the front and are not intended to be installed in cabinets.

switch, bypass isolation: A manually operated switch that works in conjunction with a transfer switch.

switch, general-use: A switch intended for use in general distribution and branch circuits. It is rated in amperes and is capable of interrupting its rated voltage.

switch, general-use snap: A type of general-use switch so constructed that it can be installed in flush device boxes or on outlet covers, or otherwise used in conjunction with wiring systems recognized by the National Electrical Code.

switching relay: An electromechanical device that will operate a set of contacts upon receiving a control signal.

switch, isolating: A switch intended for isolating an electrical circuit from the source of power. It has no interrupting rating and is intended to be operated only after the circuit has been opened by some other means.

switch, knife: A switch in which the circuit is closed by a moving blade engaging contact clips.

switch-leg: That part of a circuit run from a lighting outlet box where a luminaire or lampholder is installed down to an outlet box that contains the wall switch that turns the light or other load on or off; it is a control leg of the branch circuit.

switch loop: An electrical load is installed between the power source and the switch. The power bypasses the load through a junction box and goes to the switch. The switch delivers the power to the device.

switch, motor-circuit: A switch, rated in horsepower, capable of interrupting the maximum operating overload current of a motor having the same horsepower rating as the switch at the rated voltage. See Fig. S-14 on page 262.

symbol: A visual object used on blueprint drawings to represent a part of the electrical circuit.

symmetrical: Exhibiting symmetry.

symmetry: The correspondence in size, form, and arrangement of parts on opposite sides of a plane or line or point.

synchronism: Connected ac systems, machines, or a combination operating at the same frequency when the phase angle displacements between voltages in them are constant, or vary about a steady and stable average value.

synchronous: Simultaneous in action and in time (in phase).

ENCLOSURE	EXPLANATION
NEMA Type 1 General Purpose	To prevent accidental contact with enclosed apparatus. Suitable for application indoors where not exposed to unusual service conditions.
NEMA Type 3 Weatherproof (Weather Resistant)	Protection against specified weather hazards. Suitable for use outdoors
NEMA Type 3R Raintight	Protects against entrance of water from a rain. Suitable for general outdoor application not requiring sleetproof.
NEMA Type 4 Watertight	Designed to exclude water applied in form of hose stream. To protect against stream of water during cleaning operations, etc.
NEMA Type 5 Dusttight	Constructed so that dust will not enter the enclosed case. Being replaced in some equipment by Square D NEMA 12 Types
NEMA Type 7 Hazardous Locations A, B, C or D Class I — Air Break Letter or letters following type number indicates particular groups of hazardous locations per N.E.C.	Designed to meet application requirements of National Electrical Code for Class I, Hazardous locations (Explosive atmospheres). Circuit interruption occurs in air.
NEMA Type 9 Hazardous Locations E, F or G Class II Letter or letters following type number indicates particular groups of hazardous locations per N.E.C.	Designed to meet application requirements of National Electrical Code for Class II Hazardous Locations (Combustible Dusts, etc.).
NEMA Type 12 Industrial Use	For use in those industries where it is desired to exclude dust, lint, fibers and filings, or oil or coolant seepage.

Fig. S-14: Practical application of safety switches used for motors and other power circuits.

synchronous machine: A machine in which the average speed of normal operation is exactly proportional to the frequency of the system to which it is connected.

synchronous motor: A motor that maintains a constant speed as long as the speed of the generator supplying it remains constant.

synchronous speed: The speed of rotation of the magnetic flux produced by linking the primary winding.

synchrotron: A device for accelerating charged particles to high energies in a vacuum; the particles are guided by a changing magnetic field while they are accelerated in a closed path.

system: A region of space or quantity of matter undergoing study.

system grounding: The connecting of the system grounding conductor to earth.

T

tabulate: To arrange items or data in a table or list.

tachometer: An instrument for measuring revolutions per minute.

tachometer generator: An instrument that checks magnitude or impulses so that rpms can be properly calibrated through dial gauge readings.

tackle: The chain, rope, pulleys, or blocks used for hoisting purposes in the erection of heavy work.

tagout: A warning tag that accompanies a padlock when working on electrical panels or equipment. Generally called "lockout/tagout."

take-off: The procedure by which a listing is made of the numbers and types of electrical components and devices for an installation, taken from the electrical plans, drawings, and specs for the job.

take-up: 1) A device to pull wire or cable. 2) The process of accumulating wire or cable.

talkset: A fiber optics device that allows conversation to be held over unused fibers.

tandem extrusion: Extruding two materials, the second being applied over the first, with the two extruders being just a few inches or feet apart in the process.

tangent (geometry): A line that touches a curve at a point so that it is closer to the curve in the vicinity of the point than any other line drawn through the point; (trigonometry) in a right triangle it is the ratio of the opposite to the adjacent sides for a given angle.

tank circuit: A parallel *LC* circuit.

tank test: The immersion of a cable in water while making electrical tests; the water is used as a conducting element surrounding the cable.

tantalum electrolytic capacitor: A capacitor that has leakage current, but less than that of aluminum electrolytic capacitors; however, it is more expensive.

tap: A splice connection of a wire to another wire (such as a feeder conductor in an auxiliary gutter) where the smaller conductor runs a short distance (usually only a few feet, but could be several feet) to supply a panelboard or motor controller or switch. Also called a "tap-off," indicating that energy is being taken from one circuit or piece of equipment to supply another circuit or load; a tool that cuts or machines threads in the side of a round hole. See Fig. T-1.

Fig. T-1: Safety switch taps from an auxiliary gutter (wire trough).

tap drill: Drill used to form hole prior to placing threads in hole. The drill is the size of the root diameter of tap threads. See Fig. T-2.

tape: A relatively narrow, long, thin, flexible fabric, film or mat or combination thereof; helically applied tapes are used for cable insulation, especially at splices; for the first century the primary insulation for cables above 2kV was oil saturated paper tapes.

taper: A gradual and uniform decrease in size, as a tapered socket, a tapered shaft, a tapered shank.

tape rule: A flexible metal measuring tape that retracts into a small reel type case and can be carried on a person's belt. See Fig. T-3.

Fig. T-2: Typical tap drill.

taping (rope): A method used to stop the end of a rope from fraying.

target, sag: A visual reference used when sagging.

Note: When working around live electricity the electrician should not use this type of measuring instrument. A wood folding rule is much safer!

Fig. T-3: Tape rule is a flexible measuring tape that retracts into a small reel-type case.

task lighting: Specific lighting to supply proper illumination to task areas such as work benches; it provides more lumens than are provided by general lighting.

TC: 1) Thermocouple. 2) Time constant. 3) Timed closing.

TDR: 1) Time Delay Relay. 2) Time domain reflectometer, pulse-echo testing of cables; signal travels through cable until impedance discontinuity is encountered, then part of signal is reflected back; distance to fault can be estimated. Useful for finding faults, broken shields, or conductor.

Technical Appeal Board (TAB) of UL: A group to recommend solutions to technical differences between UL and a UL client.

Teflon®: A DuPont trade name for polytetrafluoroethylene which is used as high temperature insulation and has low dissipation factor and low relative capacitance.

telecom: Telephone communication system.

telegraphy: Telecommunication by the use of a signal code.

telemetering: Measurement with the aid of intermediate means that permits interpretation at a distance from the primary detector.

telephone: The transmission and reception of sound by electronics.

temper: A measure of the tensile strength of a conductor; indicative of the amount of annealing or cold working done to the conductor.

temperature, ambient: The temperature of the surrounding medium, such as air around a cable.

temperature, coefficient of resistance: The unit change in resistance per degree temperature change.

temperature, emergency: The temperature to which a cable can be operated for a short length of time, with some loss of useful life.

temperature humidity index: Actual temperature and humidity of a sample of air compared to air at standard conditions.

temperature, operating: The temperature at which a device is designed or rated for normal operating conditions; for cables: the maximum temperature for the conductor during normal operation.

temperature relay: A relay, set at a predetermined temperature, designed to protect other equipment or devices and not itself.

temperature rise: 1) Temperatures that rise above ambient temperature. 2) Some electrical equipment has ratings that show the allowable temperature rise the equipment may operate in.

template: A pattern used to mark out electrical devices accurately.

temporary: This single word is used to mean either "temporary service" (which a power company will give to provide electric power in a building under construction) or "temporary inspection" (the inspection that a code-enforcing agency will make of a temporary service prior to inspection of the electrical work in a building under construction). See Fig. T-4, on page 269.

tensile strength: The greatest longitudinal stress a material—such as a conductor—can withstand before rupture or failure while in service.

tension, final unloaded conductor: The tension after the conductor has been stretched for an appreciable time by loads simulating ice and wind.

tension, initial conductor: The tension prior to any external load.

tension, working: The tension that should be used for a portable cable on a power reel; it should not exceed 10% of the cable breaking strength.

term: An algebraic expression in which letters and numbers are not separated by plus and minus signs.

terminal: A device used for connecting cables. See Fig. T-5, on page 270.

termination: 1) The connection of a cable. 2) The preparation of shielded cable for connection.

Tesla: A unit of measure. 1 Tesla = 1 weber per square meter.

Tesla, Nikloa: A close friend of George Westinghouse, who argued that ac was the more practical approach to lighting and talked the Westinghouse Company into lighting the first world's fair with ac.

testlight: Light provided with test leads that is used to test or probe electrical circuits to determine if they are energized.

test, proof: Made to demonstrate that the item is in satisfactory condition.

Fig. T-4: Components for one type of temporary electrical service.

Fig. T-5: Several types of cable terminals.

test, voltage breakdown: 1) Step method—applying a multiple of rated voltage to a cable for several minutes, then increasing the applied voltage by 20% for the same period until breakdown. 2) Applying a voltage at a specified rate until breakdown.

test, voltage life: Applying a multiple of rated voltages over a long time period until breakdown; time to failure is the parameter measured.

test, volume resistivity: Measuring the resistance of a material such as the conducting jacket or conductor stress control.

test, water absorption: Determination of how much water a given volume of material will absorb in a given time period; this test is being superseded by the EMA test.

therm: Quantity of heat equivalent to 100,000 Btu.

thermal compound: A bonding agent that thermally bonds two surfaces together to improve the rate of heat transfer from one surface to the other.

thermal cutout: An overcurrent protective device containing a heater element in addition to, and affecting, a renewable fusible member that opens the circuit. It is not designed to interrupt short-circuit currents. See Fig. T-6 on page 271.

Thermistor

Fig. T-6: Heaters for thermal relays are made with different current ratings, so that within its limits any starter can be used with different size motors and still provide proper protection by selecting the size of heater that corresponds to the full-load current of the motor being used.

thermal endurance: The relationship between temperature and time of degrading insulation until failure, under specified conditions.

thermally protected (as applied to motors): When the words thermally protected appear on the nameplate of a motor or motor-compressor, it means that the motor is provided with a thermal protector designed to protect the motor from overloads.

thermal protector (as applied to motors): A protective device that is assembled as an integral part of a motor or motor-compressor and that, when properly applied, protects the motor against dangerous overheating due to overload and failure to start.

thermal relay (hot wire relay): Electrical control used to actuate a refrigeration system. This system uses a wire to convert electrical energy into heat energy.

thermal shock: Subjecting something to a rapid, large temperature change.

thermionic emission: The liberation of electrons or ions from a solid or liquid as a result of its thermal energy.

thermistor: An electronic device that makes use of the change of resistivity of a semiconductor with change in temperature.

thermocouple: A device using the Seebeck effect to measure temperature.

thermodisk defrost control: Electrical switch with bimetal disk that is controlled by electrical energy.

thermodynamics: Science that deals with the relationships between heat and mechanical energy and their interconversion.

thermoelectric: The electricity produced when heat is applied to a junction of two dissimilar metals.

thermoelectric generator: A device interaction of a heat flow and the charge carriers in an electric circuit, and that requires, for this process, the existence of a temperature difference in the electric circuit.

thermoelectric heat pump: A device that transfers thermal energy from one body to another by the direct interaction of an electrical current and the heat flow.

thermometer: Device for measuring temperatures.

thermopile: When thermocouples are connected together, a thermopile is created.

thermoplastic: Materials which, when reheated, will become pliable with no change of physical properties.

thermoset: Materials that may be molded, but when cured, undergo an irreversible chemical and physical property change.

thermostat: Device responsive to ambient temperature conditions.

thermostat wire: Same as bell wire or annunciator wire.

thermostatic expansion valve: A control valve operated by temperature and pressure within an evaporator coil, which controls the flow of refrigerant.

thermostatic valve: Valve controlled by thermostatic elements.

theta: The phase angle between phasors.

Thevenin's theorem: A theorem that reduces a complex circuit into a single source of emf, resistance, and load.

THHN: 0 to 2000 volt wire, installed in a 90°C ambient temperature or less, and in a damp or dry location.

thickness gauge: It is shaped somewhat like a pocketknife, and has blades varying in thickness by thousandths of an inch. It is used in adjusting parts with a desired amount of clearance.

threaded boss: The hub that fits on top of a weather-tight enclosure that connects conduit to the enclosure. See Fig. H-7.

threaded nipple: A threaded metal tube that connects to a threaded stud as part of a light fixture.

threaded stud: A threaded metal tube that screws into a fixture mounting strap to hold a ceiling light fixture.

threading: The cutting of screw threads, either internal or external.

three-phase circuit: A polyphase circuit of three interrelated voltages for which the phase difference is 120°; the common form of generated power.

three-way switch: A switch that is used to control a light, or set of lights, from two different points. See Fig. T-7.

throw: The number of internal operations a switch can perform.

thumper: A device used to locate faults in a cable by the release of power surges from a capacitor, characterized by the audible noise when the cable breaks down.

Fig. T-7: Operating characteristics of three-way switches. Operating either switch will change the current path, allowing the lamp to be controlled (turned ON and OFF) from either location.

thundercloud: A cloud that contains both positive and negative electrical charges and produces lightning and thunder.

THWN: 0 to 2000 volt conductor, installed in a 75°C ambient temperature or less, and in a wet or dry location.

thyratron: A gas-filled triode tube that is used in electronic control circuits.

thyristor: An electronic device that operates only in the on or off condition.

time-clock switch: Any switch that is controlled by a clock.

time constant: The length of time for a capacitor to either charge or discharge 63.2% of the available voltage.

time-delay fuse: A fuse designed to handle short electrical surges without opening the circuit.

timer: Mechanism used to control on and off times of an electrical circuit.

timer-thermostat: Thermostat control that includes a clock mechanism. Unit automatically controls room temperature and changes it according to the time of day.

tinned: Having a thin coating of pure tin or tin alloy; the coating may keep rubber from sticking or be used to enhance connection; coatings increase the resistance of the conductor, and may contribute to corrosion by electrolysis.

tip conductor: The telephone conductor that is connected to the positive terminal. See Fig. C-2.

title block: The outlined space usually in the lower right corner, or in strip form across the bottom of a drawing, containing name of company, title of drawing, scale, date, and other pertinent information.

toggle: A device having two stable states: i.e., toggle switch that is used to turn a circuit ON or OFF. A device may also be toggled from slow to fast, etc.

toggle bolt: Used for attaching articles to a hollow-tile wall. It consists of a screw with a swivel piece attached near the end. This piece may be swung into a length-wise position while the bolt is being inserted, after which it swings to a right-angle position, thus permitting "pulled up" on the bolt.

tolerance: The permissible variation from rated or assigned value.

tool pouch: A leather or heavy nylon pouch that has compartments for holding hand tools. A belt is usually included so the pouch can be fastened to the electrician's waist.

topping-off: The finishing touches put to an electrical installation; mounting plates on wall switches, receptacles, and other wiring devices; installing fixtures, etc.

toroid: A coil wound in the form of a doughnut; i.e., current transformers.

torpedo level: A level that is about 8 to 10 inches in length with a center leveling bubble that can be seen by looking at the top or sides of the level. The level has a second bubble that can be used for checking the plumb and a third bubble that is on a 45° angle to check angles with.

torque: The turning effort of a motor.

torquing: Applying a rotating force and measuring or limiting its value.

TPE: Thermal Plastic Elastomer.

TPR: Thermal Plastic Rubber.

tracer: A means of identifying cable.

track lighting: A set of adjustable lights that connect to a metal electrical carrying track on the ceiling, wall, or soffit.

trade union: An alliance of workers organized for the purpose of securing standardized privileges for all its members.

transducer: A device by means of which energy can flow from one or more media to another.

transfer switch: A device for transferring one or more load conductor connections from one power source to another.

transformer: An electromagnetic device used to 'step up' or 'step down' current or voltage. This allows the transfer of desired power.

transformer losses: Power losses in a transformer caused by eddy currents, hysteresis, and winding resistance.

transformer, potential: Designed for use in measuring high voltage; normally the secondary voltage is 120V.

transformer, power: Designed to transfer electrical power from the primary circuit to the secondary circuit(s) to 1) step-up the secondary voltage at less current or 2) step-down the secondary voltage at more current; with the voltage-current product being constant for either primary or secondary.

transformer-rectifier: Combination transformer and rectifier in which input in ac may be varied and then rectified into dc.

transformer, safety isolation: Inserted to provide a non-grounded power supply such that a grounded person accidentally coming in contact with the secondary circuit will not be electrocuted.

transformer, vault-type: Suitable for occasional submerged operation in water.

transient: 1) Lasting only a short time; existing briefly; temporary. 2) A temporary component of current existing in a circuit during adjustment to a changed load, different source voltage, or line impulse.

transistor: An active semiconductor device usually with three or more terminals. See Fig. T-8.

transmission: Transfer of electric energy from one location to another through conductors or by radiation or induction fields.

Fig. T-8: Transistors used in an electric circuit.

transmission line: A long electrical circuit.

transposition: Interchanging position of conductors to neutralize interference.

traveler: 1) A pulley complete with suspension armor frame to be attached to overhead line structures during stringing. 2) *Travelers* are the two conductors between two three-way switches. See Fig. T-7, on page 273.

triac: A full wave device that combines two SCRs to get the bidirectional function.

triangle: A closed three-sided figure.

trickle charge: A rectifier changing ac to dc, and delivering same to a storage battery for a certain period of time, usually at a very minute rate.

trigger pulses: The voltage or current necessary to activate a gate, base, or input to an electronic device.

trigonometry: The understanding of the properties and applications of trigonomic functions.

trigonometric functions: Relationships between the angles and sides of a right triangle (i.e., sine, cosine, and tangent).

trim: The front cover assembly for a panel, covering all live terminals and the wires in the gutters, but providing openings for the fuse cutouts or circuit breakers mounted in the panel; may include the door for the panel and also a lock.

triode: A three-electrode electron tube containing an anode, a cathode, and a control electrode.

triplex: Three cables twisted together.

tri-stand pipe vise: A three-legged vise that folds to make it easier to carry and to store. It has either a chain or a screw jack to hold the pipe in place while it is being worked on. Also called a tri-vise.

trolley wire: Solid conductor designed to resist wear due to rolling or sliding current pickup trolleys.

troubleshooting: A systematic method of defining the problem, gathering information about the problem, analyzing the information gathered about the problem, then making a decision on how to correct the problem and, finally, receiving feedback. Did the system of analysis work? If not why not?

trough: Another name for an "auxiliary gutter," which is a sheet metal enclosure of rectangular cross-section, used to supplement wiring spaces at meter centers, distribution centers, switchboards, and similar points in wiring systems where splices or taps are made to circuit conductors. The single word "gutter" is also used to refer to this type of enclosure.

true RMS meter: An ammeter that responds only to true RMS amps and voltage.

trunk feeder: A feeder connecting two generating stations or a generating station and an important substation.

trussed: Framed structural pieces consisting of triangles in a single plane for supporting loads over spans.

truth table: A method of indicating contact position on a selector switch.

tub: An expression sometimes used to refer to large panelboards or control center cabinets, in particular the box-like enclosure without the front trim.

tube: A long, hollow product having uniform wall thickness and uniform cross-section.

tungsten lamp: A type of incandescent lamp having a filament of fine tungsten wire. See Fig. T-9 on page 279.

tuning: The adjustment of a circuit or system to secure optimum performance.

turn: The basic coil element that forms a single conducting loop comprised of one insulated conductor.

turn ratio: The ratio between the number of turns between windings in a transformer, normally primary to secondary; however, for current transformers it is the ratio of the secondary to primary. See Fig. T-10 on page 279.

turns-per-volt ratio: See *turn ratio*.

TVSS: Transient Voltage Surge Suppression. This is a surge protector.

TW: A type of wire insulation used on 0 to 2000 volt wire, installed in a 60°C ambient temperature or less, and in a dry or wet location.

Fig. T-9: The quartz-iodine tungsten-filament lamp is basically an incandescent lamp since light is produced from the incandescence of its coiled tungsten filament. However, the lamp envelope, made of quartz, is filled with iodine vapor that prevents the evaporation of the tungsten filament.

twisted cable: A cable consisting of two or more conductors twisted together; often referred to as bell or doorbell wire. Those recently manufactured normally have a non-metallic jacket covering the conductors. A ground or bonding wire is sometimes integrated within the jacket. Currently used for low-voltage control wiring and similar applications.

Fig. T-10: Since there are 1000 turns in the primary of this transformer and only 100 turns in the secondary, the turn ratio is 10:1.

twists (rope): To wind two or more strands around each other in the process of making rope.

twist test: A test to grade round material for processibility into conductors.

two-phase: A polyphase ac circuit having two interrelated voltages.

U

U bolt: A bolt shaped like the letter *U* threaded at both ends; used mainly to secure conduit in place.

UF cable: Electrical cable suited for underground burial.

Ufer ground: Bare copper conductor used when ordinary soil conditions cannot be relied upon for grounding purposes.

UHF (Ultra High Frequency): 300 MHz to 3 GHz.

UL (Underwriters Laboratories): A nationally known laboratory for testing a product's performance with safety to the user being the prime consideration; UL is an independent organization, not controlled by any manufacturer, the best known laboratory for testing electrical products.

UL listed: An item that has been tested and approved to be used as the manufacturer has intended.

UL recognized: A product that is not to be used by itself.

ultimate strength: The highest unit stress that can be sustained; it occurs just at or just before rupture.

ultrasonic: Sounds having frequencies higher than 20 KHz, which is at the upper limit of human hearing.

ultrasonic cleaning: Immersion cleaning aided by ultrasonic waves that cause microagitation.

ultrasonic detector: A device that detects ultrasonic noise such as that produced by a corona or leaking gas.

ultraviolet: Radiant energy within the wave length range 10 to 380 nanometers. This energy is invisible, filtered by glass, causes teeth to glow, and causes suntan.

undercurrent: Less than normal operating current.

underground cable: A single or multiple conductor cable sheathed in lead or other waterproof materials for installation below grade. Underground cable used to feed a building's electric service is shown in Fig. U-1.

Fig. U-1: An underground electric service (service lateral) fed with direct-burial underground conductors.

underground service: When the service cable is buried underground from the building's point of connection to the utility companies' point of connection.

undervoltage: Less than normal operating voltage.

undervoltage protection: A device that opens the circuit if undervoltage occurs to prevent a starter coil from burning out.

undervoltage release: A device that opens the circuit if undervoltage occurs but does not prevent the return of voltage to the main circuit.

underwriter: Name given to a representative of an inspection organization who examines electrical installations for life and fire hazards.

underwriter's knot: Special knot that holds wires in the proper place on a plug or lamp. See Fig. U-2.

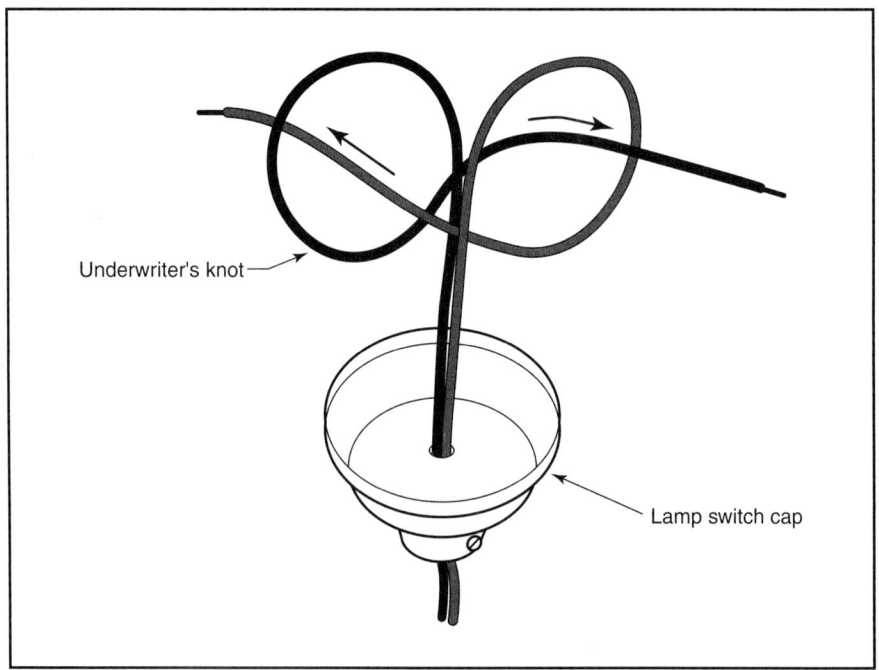

Fig. U-2: Underwriter's knot is a special knot used to hold electric wires in their proper place on a plug or lamp.

ungrounded: Not connected to the earth intentionally.

unijunction transistor: A transistor that is not used as an amplifier but only used in timing and control applications.

union: A coupling or connection for threaded conduit.

unit magnetic pole: One that will repel an equal and like pole with a force of one dyne at a distance of 1 cm. A *dyne* is the force which, acting upon a mass of 1 gram during 1 second, gives the mass a velocity of 1 cm per second.

unit of magnetic flux: The total number of lines of force set up in a magnetic substance. Treated as a magnetic current flowing in a magnetic circuit.

unit of magnetic intensity: Magnetomotive force. Magnetic pressure that drives lines of force through a magnetic circuit.

universal joint: A type of coupling that permits the free rotation of two shafts whose axes are not in a straight line.

Universal motor: A motor designed to operate on either ac or dc at about the same speed and output using either power source. See Fig. U-3.

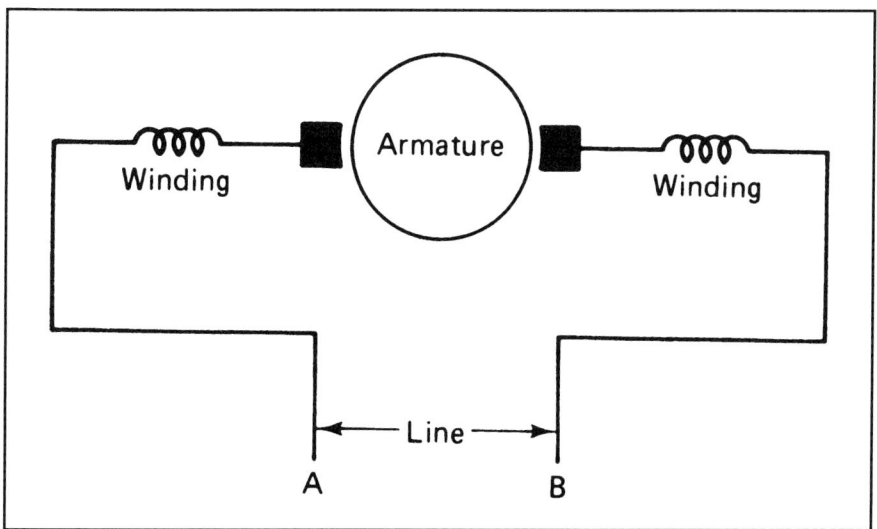

Fig. U-3: Universal motor wiring diagram with the principal parts.

Urethane foam: Type of insulation that is foamed in between inner and outer walls.

URD (Underground Residential Distribution): A single-phase cable usually consisting of an insulated conductor having a bare concentric neutral.

useful static discharge: Static electricity is used in electronic filters and in spray painting, and has become very useful in many ways.

U-shaped lamp: A U-shaped fluorescent fixture.

utility: A public service such as a telephone or electric company. A power company furnishing electric service to users is a utility.

utilization equipment: Equipment that uses electric energy for mechanical, chemical, heating, lighting, or other useful purposes. Examples would be electric motors, large and small appliances, electric heaters, air-conditioning equipment, and the like.

V

va: Volts times Amps, which equals Watts.

vacuum: Reduction in pressure below atmospheric pressure.

vacuum fish tape: A power tool that creates a strong vacuum that pulls a piston through a conduit. A string line dispenser is part of the tool.

vacuum pump: Special high-efficiency compressor used for creating high vacuums for testing or drying purposes.

vacuum switch: A switch with contacts in an evacuated enclosure.

vacuum tube: A device, usually in the form of a tube of glass or metal, from which the air has been exhausted and is provided with two or more elements. This sealed glass enclosure has two or more electrodes between which conduction of electricity can take place.

valance: A board that is attached to a wall, ceiling, or cabinet to direct light to a certain area. See Fig. I-2, page 126.

valance lighting: Fluorescent lighting that is installed and shielded by a valance.

valence electrons: Electrons that occupy the outermost shell of an atom.

285

valence shell: The outermost orbit of an atom.

valley: 1) The gutter or angle formed by the meeting of two roof slopes. The space between vault ridges viewed from above.

valve: Device used for controlling fluid flow.

valve, expansion: Type of refrigerant control that maintains a pressure difference between high-side and low-side pressure in a refrigerating mechanism. The valve operates by pressure in the low or suction side.

valve, solenoid: Valve actuated by magnetic action by means of an electrically energized coil.

vapor: Word usually used to denote vaporized refrigerant rather than gas.

vapor barrier: Thin plastic or metal foil sheet used in air-conditioned structures to prevent water vapor from penetrating insulating material.

vaporize: To convert into a state of gas or vapor.

vapor lock: Condition where liquid is trapped in line because of a bend or improper installation that prevents the vapor from flowing.

vapor-safe: Constructed so that a device or apparatus may be operated without hazard to its surroundings in hazardous areas in aircraft.

vapor, saturated: A vapor condition that will result in condensation into liquid droplets as vapor temperature is reduced.

vapor-tight: So enclosed that vapor will not enter.

variable: A quantity that may change in value while the others remain constant.

variable resistance: A resistance that can be adjusted or changed to different values.

variable speed drive: A motor having an integral coupling device that permits the output speed of the unit to be easily varied through a continuous range.

varistor: Designed to protect circuits from sudden increases from the source voltage. Also called voltage-sensitive or voltage-dependent resistors.

varnish: An insulation that is applied as a liquid and is quite thin.

VARs An electrical measurement that measures reactive power.

varying duty: That demand which requires operation of loads of varying quantity for intervals of time, both of which may be subject to wide variation.

V Block: V-shaped groove in a metal block used to hold a shaft.

VD: Voltage Drop.

vector: A mathematical term expressing magnitude and direction.

ventilation: Circulation of air; system or means of providing fresh air.

vermiculite: Lightweight inert material resulting from expansion of mica granules at high temperatures; used as an aggregate in plaster.

vernier: An auxiliary scale permitting measurements more precise than the main scale.

vertical: Plumb, perpendicular, upright.

vibrating bell: An electric device having a clapper or hammer that strikes a bell rapidly when an electric current flows through it. It operates on the principle of electromagnetic attraction.

vibrating-reed meter: A meter that is used primarily to read low frequencies in power systems.

vibrator coil: An induction coil so constructed that the magnetism of the core operates the make and break or vibrator of the primary circuit.

viscosity: Internal friction or resistance to flow of a liquid; the constant ratio of shearing stress to rate of shear.

vise: A mechanical device for holding a piece of wood or metal while it is being worked on. It consists essentially of two jaws, one fixed and one movable, the movable jaw being operated by a screw by means of which clamping action is secured. Some vises are equipped with pipe jaws for securing pipe or conduit.

voice fire alarm system: A fire alarm system that has loudspeakers that amplify recorded warnings or live voices.

volatile flammable liquid: A flammable liquid having a flash point below 38°C or whose temperature is above its flash point.

volt: The derived SI unit for voltage: one volt equals one watt per ampere.

voltage: The electrical property that provides the energy for current flow; the ratio of the work done to the value of the charge moved when a charge is moved between two points against electrical forces.

voltage, breakdown: The minimum voltage required to break down an insulation's resistance, allowing a current flow through the insulation, normally at a point.

voltage, contact: A small voltage established whenever two conductors of different materials are brought into contact; due to the difference in work functions or the ease with which electrons can cross the surface boundary in the two directions.

voltage divider: A network consisting of impedance elements connected in series to which a voltage is applied and from which one or more voltages can be obtained across any portion of the network.

voltage drop: The loss in voltage between the input to a device and the output from a device due to the internal impedance or resistance of the device. In all electrical systems, the conductors should be sized so that the voltage drop never exceeds 3 percent for power, heating, and lighting loads or combinations of these. Furthermore, the maximum total voltage drop for conductors for feeders and branch circuits combined should never exceed 5 percent.

The voltage drop in any two-wire, single-phase circuit consisting of mostly resistance-type loads, with the inductance negligible, may be found by the following equation:

$$VD + \frac{2K \times L \times I}{CM}$$

where

VD = drop in circuit voltage
L = length of conductor

I = current in the circuit
CM = area of conductor in circular mils
K = resistivity of conductor metal, that is, 11 for copper and 18 for aluminum

voltage drop dissipation: When heat is produced by voltage drop in a switching device, the heat is dissipated by a method of heat sinks.

voltage, EHV (Extra High Voltage): 230 to 765 kV.

voltage, induced: A voltage produced in a conductor by a change in magnetic flux linking that path.

voltage rating of a cable: Phase-to-phase ac voltage when energized by a balanced three-phase circuit having a solidly grounded neutral.

voltage regulator: A device to decrease voltage fluctuations to loads.

voltage relay: A relay that operates either at an overvoltage or undervoltage and has a preset voltage operation.

voltage, signal: Voltages to 50 V.

voltage tester: An instrument that tests whether voltage is present or not.

voltage to ground: The voltage between an energized conductor and earth.

voltage, UHV (Ultra-High Voltage): 765+ kV.

voltaic cell: A cell whose electrical charge is created by suspending two dissimilar elements in an acid solution.

volt-ampere: Volt-amps = watts.

voltanic cell: Primary cell. Name given to the cell first discovered by Volta. Sometimes called a galvanic cell. It is a cell in which two dissimilar metals are immersed in a solution that is capable of acting chemically more upon one than on the other, to produce a difference of potential (voltage) across the metals.

voltmeter: An instrument for measuring voltage. See Fig. V-1, on page 290.

VOM (Volt-Ohm-Multimeter): An electrical test instrument commonly used to test voltage, current, resistance, and continuity.

Fig. V-1: One type of testing meter that reads voltage, amperage, and resistance.

vortex tube: Mechanism for cooling or refrigerating that accomplishes a cooling effect by releasing compressed air through a specially designed opening. Air expands in a rapidly spiraling column of air that separates slow-moving molecules (cool) from fast-moving molecules (hot).

vulcanize: To cure by chemical reaction that induces extensive changes in the physical properties of a rubber or plastic, brought about by reacting it with sulphur and/or other suitable agents; the changes in physical properties include decreased plastic flow, reduced surface tackiness, increased elasticity, much greater tensile strength, and considerably less solubility; the process being hastened by heat and pressure; the method used for curing thermosetting materials—rubbers, XLP, etc.

VW-1: A UL rating given to single conductor cables regarding their flame resistant properties; formerly FR-1.

wall: The thickness of insulation or jacket of cable.

wallboard: A general term applied to any of the many building boards used in place of plaster on interior walls and ceilings.

wallboard anchor: An expandable anchor to secure screws in wallboard.

wall, fire: A dividing wall for the purpose of checking the spread of fire from one part of a building to another.

wall plate: A flush-mounted receptacle or switch plate used to cover the outlet box and wiring device.

wall socket: An electric outlet located in or on the wall for the purpose of providing a source of current.

Warren motor: A motor that is constructed with a laminated core and a single core.

washer: A small, flat, perforated disk, used to secure the tightness of a joint, screw, etc.

waterblocked cable: A multiconductor cable having interstices filled to prevent water flow or wicking.

water-cooled condenser: Condensing unit that is cooled through the use of water.

waterproof: So constructed that moisture will not interfere with successful operation.

water table: A projection of the wall at or near the grade line of a building. It is used to turn the water away from the foundation wall.

watertight: So constructed that water will not enter.

watt: The derived SI unit for power, radiant flux; one watt equals one joule per second.

wattage rating: An appliance or lighting power rating that has been established by the manufacturer.

watt-hour: The number of watts used in one hour.

watt-hour meter: A meter that measures and registers the integral, with respect to time, of the active power in a circuit. A typical watt-hour meter consists of a combination of coils, conductors, and gears—all encased in a housing as shown in Fig. W-1. The coils are constructed on the same principle as a split-phase induction motor, in that the stationary current coil and the voltage coil are placed so that they produce a rotating magnetic field. The disc near the center of the meter is exposed to the rotating magnetic field. The torque applied to the disc is proportional to the power in the circuit, and the braking action of the eddy currents in the disc makes the speed of rotation proportional to the rate at which the power is consumed. The disc, through a train of gears, moves the pointers on the register dials to record the amount of power used directly in kilowatt hours (kWh).

Fig. W-1: A typical watt-hour meter.

Most watt-hour meters now have five dials, as shown in Fig. W-1. The dial farthest to the right on the meter counts the kilowatt hours singly. The second dial from the right counts by tens, the third dial by hundreds, the fourth dial from the right by thousands, and the left-hand dial by ten-thousands.

wattmeter: An instrument for measuring the magnitude of the active power in a circuit.

watt-second: Unit of energy (1 joule).

wave: A disturbance that is a function of time or space or both.

waveform: The geometrical shape as obtained by displaying a characteristic of the wave as a function of some variable when plotted over one primitive period.

wavelength: The distance measured along the direction of propagation between two points that are in phase on adjacent waves.

wave wound armature: An armature that is connected in series and used in equipment that requires high voltage and low current.

weatherhead: The conduit fitting at a conduit that is used to allow conductor entry, but that prevents weather entry. See Fig. W-2.

weatherproof: So constructed or protected that exposure to the weather will not interfere with successful operation.

web: An open braid; central portion of an I beam.

weber: The base unit of magnetic flux.

weight correction factor: The correction factor necessitated by uneven geometric forces placed on cables in conduits; used in computing pulling tensions and sidewall loads.

Fig. W-2: One type of weatherhead used for electric services and other outdoor applications.

weld: The joining of materials by fusion or recrystallization across the interface of those materials using heat or pressure or both, with or without filler material.

welding cables: Flexible cable used for leads to the rod holders of arc-welders, usually consisting of size 4/0 flexible copper conductors.

welding transformer: A step-down transformer used to produce sufficient instantaneous current to fuse the metals, in contact, through which it is flowing.

Western Union splice: The electrical connection made by paralleling the bared ends of two conductors and then twisting these bared ends, each around the other. See Fig. W-3.

Fig. W-3: Western Union splice.

wet bulb: Device used in measurement of relative humidity.

wet cell battery: A battery having a liquid electrolyte.

wet locations: Exposed to weather or water spray or buried.

wet niche fixture: A swimming pool light fixture in which the lens is not less than 18 inches below the normal water line. It is designed to be relamped from the front. See Fig. N-4.

Wheatstone bridge: A method of measuring resistances by the proportion existing between the resistance of three known adjustable resistances and the one to be found, all forming the arms of the bridge. See Fig. W-4.

whipping (rope): A method to prevent the end of a rope from fraying.

Whitney punch: A tool designed to punch small holes in light metal.

winch: A machine for hoisting or hauling.

winding: An assembly of coils designed to act in concert to produce a magnetic flux field or to link a flux field.

wing nut: A form of nut that is tightened or loosened by two thin flat wings extending from opposite sides.

Wire cutter

Fig. W-4: Schematic diagram of a Wheatstone bridge. The coil or line for which the resistance is to be measured is connected as R_x in the bridge circuit. The resistance circuits, R_A, R_B, and R_C are known resistances and are called bridge arms.

wiped joint: A joint wherein filler metal is applied in liquid form and distributed by mechanical action.

wire: A slender rod or filament of drawn metal; the term may also refer to insulated wire.

wire bar: A cast shape that has a square cross section with tapered ends.

wire brush: A hand brush fitted with wire or thin strips of steel instead of bristles. Used for removing rust, dirt, or foreign matter from a surface.

wire, building: That class of wire and cable, usually rated at 600 V, which is normally used for interior wire of buildings.

wire, covered: A wire having a covering not in conformance with *NEC*® standards for material or thickness.

wire cutter: A tool that leaves a blunt edge when it cuts electrical cable.

wire dispenser: A tool that can transport and hold rolls of electrical wire and cable.

wire drawing: The process by which wire is made; as by drawing metal through a hole in a steel plate.

wire, hookup: Insulated wire for low voltage and current in electronic circuits.

wireman's pliers: Another name for lineman's pliers.

wire nut: A plastic device with a metal interior that covers a wire splice and keeps the splice from coming loose. It also keeps anyone from coming in contact with a live junction wire.

wire pulling lubricant: A special slick-type material that is applied to the outside of the wire covering to keep it from creating a resistance when pulling the wire through conduit.

wire, resistance: Wire having appreciable resistance; used in heating applications such as electric toasters, heaters, etc.

wire stripper: A tool designed to strip the insulation off an electrical wire without damaging the wire.

wire wound resistors: Fixed resistors that have a piece of resistive wire wound around a ceramic rod.

wiring plan: The part of a complete set of blueprints that shows the location of electrical equipment and the overall scheme of the electrical apparatus.

withstand rating: The highest short current the equipment can withstand as established by the manufacturer.

WM: Wattmeter.

work: Force times distance: pound-feet.

work function: The minimum energy required to remove an electron from the Fermi level of a material into field-free space; units—electron volts.

work hardening: Hardening and embrittlement of metal due to cold working.

working drawing: A drawing containing all dimensions and instructions necessary for successfully carrying a job to completion. See Fig. W-5.

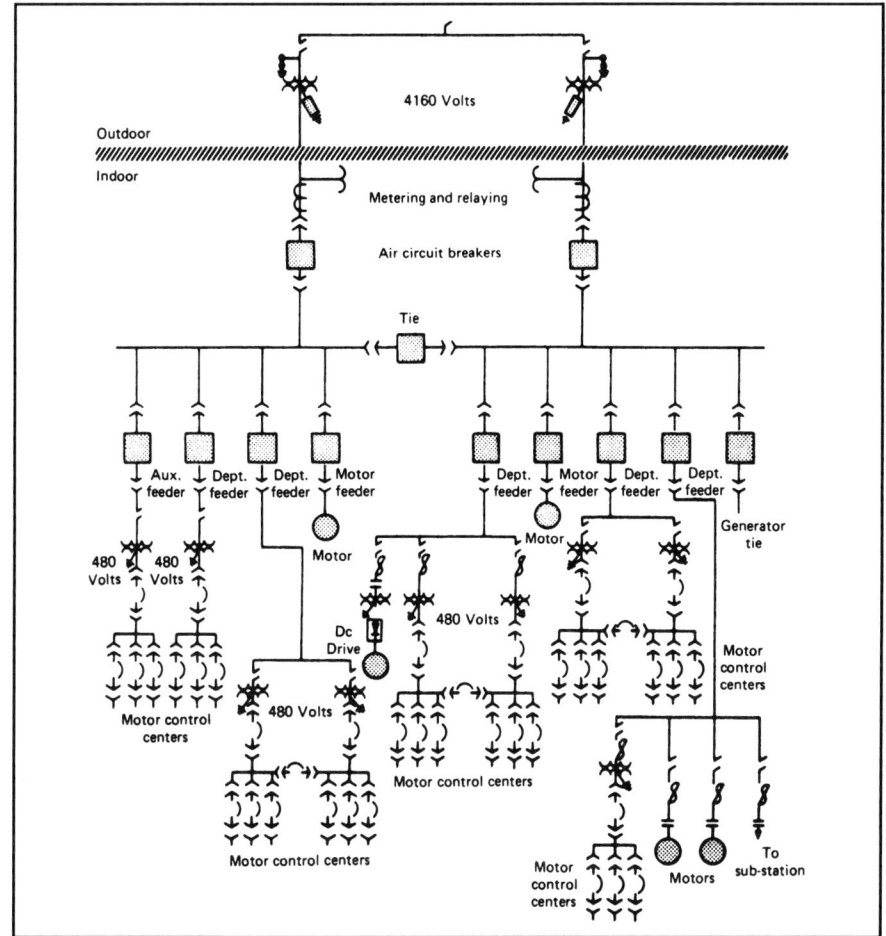

Fig. W-5: A sample drawing sheet from a set of electrical working drawings.

working loads: The expected current load of a circuit.

WOW (pipe bending): An improperly made pipe bend.

wrench: A tool for exerting a twisting strain, as in tightening a nut or bolt.

wye connection: Three-phase transformer connection wherein all three phases are connected at a central point, forming a "Y" configuration. See Y-Delta connection and Y-Y connection on pages 300 and 301.

X

X_0: The neutral connection on a transformer.

xenon: A chemical element Xe, atomic number 54. It is a member of the family of noble gases, group O in the periodic table. Its main use is to fill a type of flashbulb employed in photography and called electronic speed light.

Xmfr: Abbreviation for transformer. Used mainly on electrical drawings. See Fig. X-1.

x-ray: 1) Penetrating short wavelength electromagnetic radiation created by electron bombardment in high-voltage apparatus; produce ionization when the rays strike certain materials. The rays pass through most objects as though they were transparent. 2) A popular name for Roentgen rays. A form of radiant energy sent out when the cathode rays of a Crookes tube strike upon the opposite walls of a tube or upon any object in the tube.

x-unit: A unit of length, approximately 10^{-11} cm, formerly used for measuring x-ray wavelengths and crystal dimensions.

xylene: One of a group of three isometric aromatic hydrocarbons. The three isometric sylenes and ethylbenzene all have in common a molecular weight of 106.2 and the simplified formula, C_8H_{10}.

Fig. X-1: A wiring diagram of transformer connections showing "Xmfr" to designate the transformers to save drafting time.

Originally designated "coal tar" hydrocarbons, these compounds have, since World War II, been almost exclusively produced from petroleum by hydroforming or catalytic reforming of appropriate naphtha fractions.

Y

yardage: Relates to cubic yards of earth excavated.

yarns (rope): Coarse fibers woven into strands to make rope.

Y-Delta Connection: A three-phase transformer connection wherein the primary phases are connected at a central point and the secondary phases are connected in series with each other. See Fig. Y-1.

Fig. Y-1: Y-Delta transformer connection for light and power.

yield point: That unit stress at which the specimen begins to stretch without increase in the load.

yield strength: 1) The point at which a substance changes from elastic to viscous. 2) The load required to produce a permanent stretch or elongation of a material.

yield value: The lowest pressure at which a plastic will flow. Below this pressure the plastic behaves as an elastic solid; above this pressure as a viscous liquid.

yoke: 1) The ferromagnetic ring or cylinder that holds the pole pieces of a dynamo-type generator and acts as a part of the magnetic circuit. 2) The system of coils used for magnetic deflection of the electron beam in cathode-ray tubes, such as television picture tubes.

Y-Y Connection: A three-phase transformer connection, whereas the three primary phases are connected at a central point. The three secondary phases are also connected at a central point. See Fig. Y-2.

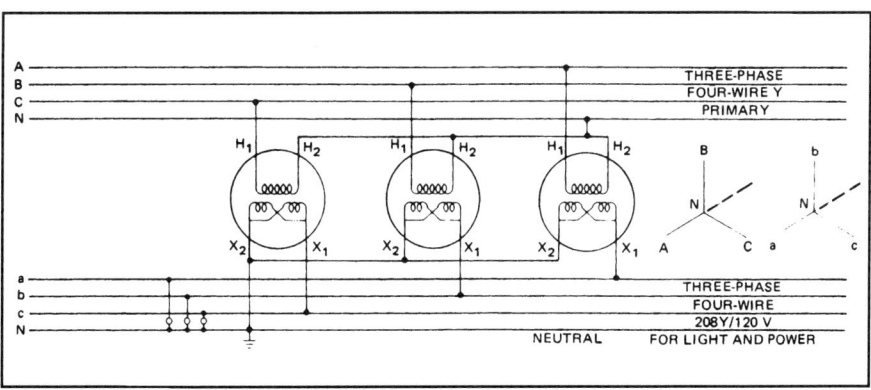

Fig. Y-2: Y-Y transformer connection for light and power.

Z

Zener diode: Works differently than a rectifier; it operates in the reverse breakdown region. See Fig. Z-1.

Zener region: Current flowing through a Zener diode in reverse direction will enter the Zener breakdown region.

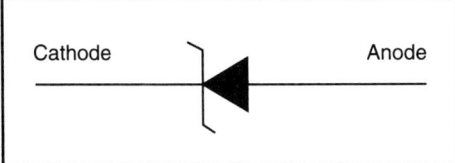

Fig. Z-1: Zener diode symbol. It operates in a reverse breakdown region.

zero: The numeral 0; a cipher; the lowest point. For example, "the voltmeter shows a reading of zero," meaning no voltage is present.

zero adjust: The mechanical adjustment of a meter to adjust the pointer back to zero.

zero potential: No voltage, no current.

zero switching: The nature of certain relays that allows current to continue to flow until the ac waveform returns to zero.

zig-zag connection: A connection of polyphase windings in which each branch generates phase-displaced voltages.

zinc: A bluish-white metallic element with a symbol of Zn, the atomic number 30, and the atomic weight 65.38. Zinc is familiar as the negative electrode material in dry cells and as a protective coating for some sheet metals used in electronics.

zinc-alkaline battery: A relatively new type of storage battery introduced some years ago. Due to its high cost, it has gained very little commercial appeal. However, space projects are utilizing this type of battery due to its greater power output compared to its size and weight. Compared with the conventional lead-acid cell, the zinc cell has a watt-hour output per pound of cell over five times the output of the lead cell, while its watt-hour output per cubic inch of cell is three to four times as much.

zinc-carbon battery: A primary cell in which the negative electrode is zinc and the positive electrode is carbon, and which may be either wet or dry. See Fig. Z-2.

zinc chloride (ZnCl$_2$): A white deliquescent salt, obtained by the solution of zinc or zinc oxide, in hydrochloric acid, or by burning zinc in chlorine. Used as a soldering flux.

zinc-copper battery: The action of the battery is such that the electrons leave the negative zinc electrode, flow to the load, and then flow on to the positive copper electrode.

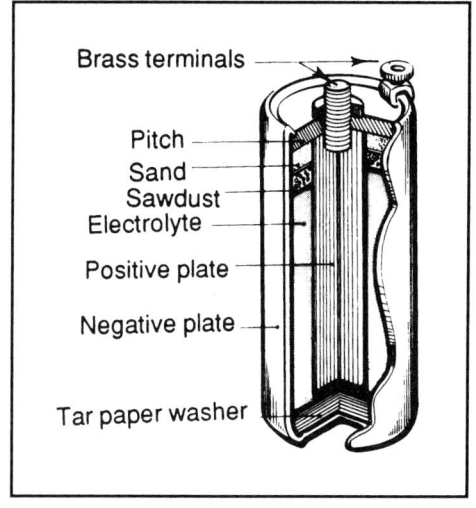

Fig. Z-2: Principal parts of a zinc-carbon battery.

zip cord: A small cord that is built into the insulating coverings of wire. The cord easily splits the covering when it is pulled.

zonal cavity method: A calculation method used to determine the amount of illumination produced by a given installation of lighting fixtures in a particular room or other indoor area.

The zonal cavity method of calculating average illumination levels assumes each room or area to consist of the following three separate cavities: ceiling cavity, room cavity, and floor cavity.

Figure Z-3 shows that the *ceiling cavity* extends from the lighting fixture plane upward to the ceiling. The *floor cavity* extends from the work plane downward to the floor, while the *room cavity* is the space between the lighting fixture plane and the working plane.

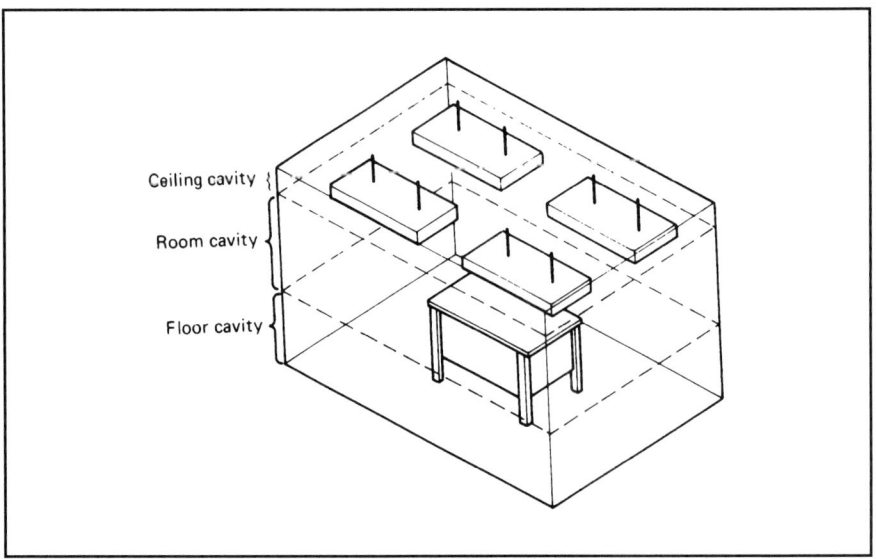

Fig. Z-3: The three room cavities as applied to the zonal cavity method of lighting calculations.

If the lighting fixtures are recessed or surface-mounted on the ceiling, there will be no ceiling cavity and the ceiling cavity reflectance will be equal to the actual ceiling reflectance. Similarly, if the work plane is at floor level, there will be no floor cavity and the floor cavity reflectance will be equal to the actual floor reflectance. The geometric proportions of these spaces become the "cavity ratios."

Cavity ratio: Rooms are classified according to shape by *ten* cavity ratio numbers. The basic formula for obtaining cavity ratios in rectangular-shaped rooms is:

$$\text{Cavity Ratio} = \frac{5 \times \text{Height}(\text{Length} + \text{Width})}{\text{Length} \times \text{Width}}$$

where height is the height of the cavity under consideration—that is, ceiling, floor, or room cavity.

The zonal cavity method of making lighting calculations provides greater flexibility with increased accuracy over any previous method of determining average illumination levels.

There are five key steps in using the zonal-cavity method:

- Determine the required level of illumination.
- Determine the coefficient of utilization.
- Determine the maintenance factor.
- Calculate the number of lamps and lighting fixtures required.
- Determine the location of the lighting fixtures.

The general applications of the zonal cavity method are to determine the number of lighting fixtures that are required to produce a given lighting level in footcandles and to determine what lighting level will be produced by a given number of lighting fixtures.

zone coded system: An audible system that identifies the zone where a fire is detected and not the specific area.

zone control: A method that controls different areas of heating and cooling in an individual's environment.

zones: A defined area within fire alarm protected premises.